THE OXYGEN ADVANTAGE

Also by Patrick McKeown

Asthma-Free Naturally

Anxiety Free: Stop Worrying and Quieten Your Mind

Sleep with Buteyko: Stop Snoring, Sleep Apnoea and Insomnia

Buteyko Meets Dr. Mew: Buteyko Method for Children and Teenagers

Close Your Mouth

THE OXYGEN ADVANTAGE

SIMPLE, SCIENTIFICALLY PROVEN BREATHING TECHNIQUES TO HELP YOU BECOME HEALTHIER, SLIMMER, FASTER, AND FITTER

PATRICK McKEOWN

wm
WILLIAM MORROW
An Imprint of HarperCollins*Publishers*

This book is written as a source of information only. The information contained in this book should by no means be considered a substitute for the advice of a qualified medical professional, who should always be consulted before beginning any new diet, exercise, or other health program.

Some names, identities, and circumstances have been changed in order to protect the anonymity of the various individuals involved.

All efforts have been made to ensure the accuracy of the information contained in this book as of the date published. The author and the publisher expressly disclaim responsibility for any adverse effects arising from the use or application of the information contained herein.

HarperCollins books may be purchased for educational, business, or sales promotional use. For information please e-mail the Special Markets Department at SPsales@harpercollins.com.

A hardcover edition of this book was published in 2015 by William Morrow, an imprint of HarperCollins Publishers.

FIRST WILLIAM MORROW PAPERBACK EDITION PUBLISHED 2016.

Designed by Diahann Sturge
Illustrations by Bex Burgess

Library of Congress Cataloging-in-Publication Data has been applied for.

ISBN 978-0-06-234947-7

23 24 25 26 27 LBC 21 20 19 18 17

This book is dedicated to all my students and readers who continue to graciously spread awareness of this work. To my late father, Patrick, who encouraged me to see things differently. And to my mother, Teresa; wife, Sinead; and daughter, Lauren—thank you for your beautiful smiles.

Contents

Some Important Guidelines Before We Begin

Although the Oxygen Advantage program is perfectly safe for the vast majority of people, part of the program includes powerful exercises that simulate high-altitude training and would be similar to performing high-intensity exercise. Just as high-intensity exercise is suited only to individuals with reasonably good health and fitness, those with any medical issues should refrain from practicing any of the exercises that simulate high-altitude training. (These exercises include the Nose Unblocking Exercise and any that simulate high-altitude training.)

If you are pregnant, this program is not suitable. For those with high blood pressure, cardiovascular disease, type 1 diabetes, kidney disease, depression, or cancer, it is advisable to practice only nasal breathing and the more gentle exercises, including "Breathing Recovery" and "Breathe Light to Breathe Right," during rest and physical activity until these conditions are resolved.

If you have any medical issues then you should follow this program only with the consent of your medical practitioner. For further information, please visit OxygenAdvantage.com.

It isn't the mountains ahead to climb that wear you out;
it's the pebble in your shoe.

—Muhammad Ali

Foreword

by Dr. Joseph Mercola

It has been well documented that those who live at higher altitudes tend to live longer. The precise mechanism behind this is not known and could be a result of several factors. However, one of the leading candidates for this explanation is a reduced pressure of oxygen at higher altitudes.

Research is very clear that calorie restriction extends life span. But another nutrient many of us don't frequently consider is oxygen. Just as excess calories can cause metabolic damage, excess oxygen can also prematurely damage your tissues through the generation of excess free radicals. These are highly reactive and destructive molecules that cause damage to the fats in your cell membranes, proteins, and DNA. Free radicals are generated by the normal breakdown of oxygen during metabolism. We all create a certain amount of free radicals through the very act of breathing, and incorporating breathing exercises designed to maintain a healthy breathing volume seems to be an effective strategy to keep your oxygen at an optimum level, and thus minimize free radical damage.

Additionally, altitude training is a tactic many elite endurance athletes use to gain a competitive edge. One way of tapping into your body's natural resources is to purposefully expose yourself to reduced oxygen intake for a short period of time. This will improve your blood's oxygen-carrying capacity and also increases the maximum volume of oxygen that an athlete can use, known as your VO_2 max.

Of course, most of us live our lives close to sea level and do not achieve this benefit. But there are simple strategies that will allow you to access the benefits of living at a high altitude with reduced oxygen intake: keeping your mouth closed while you are breathing and practicing the various exercises outlined in this book. This is a challenge during intense exercise due to air hunger, but this is when most of the benefit actually occurs. I have personally implemented the information in *The Oxygen Advantage* during my high-intensity workouts. It took me several weeks to make the transition to breathing through my nose the entire time, but once accomplished, breathing became a far more efficient process for me.

Many may know that I am a major fan of using simple, inexpensive lifestyle changes to avoid expensive and dangerous medications and surgery. The strategies in *The Oxygen Advantage* are tools that I believe should be included in your health habit arsenal. There simply are no downsides that I can identify, and there are massive upsides. I personally use this program and would strongly encourage you to apply it in your life so you can reap the rewards.

INTRODUCTION
Do More with Less

We can live without food for weeks and water for days, but air for just a few brief minutes. While we spend a great deal of time and attention on what we eat and drink, we pay practically no attention to the air we breathe. It is common knowledge that our daily consumption of food and water must be of a certain quality and quantity. Too much or too little spells trouble. We also recognize the importance of breathing good-quality air, but what about the *quantity*? How much air should we breathe for optimum health? Wouldn't it be fair to surmise that air, even more important than food or water for human survival, must also meet basic requirements?

The quantity of the air you breathe has the potential to transform everything you thought you knew about your body, your health, and your performance, whether you're a "pre-athlete" just trying to get off the couch, a weekend warrior running an occasional 10K, or a professional athlete in need of a game-changing edge over your competition.

You may wonder what I mean by quantity. After all, air isn't exactly something you can binge on at the kitchen table late at night, or take too many swigs of over the weekend. But what if, in a certain sense, it was? What if healthy breathing habits were just as important as healthy eating habits in fostering maximum fitness—or, in fact, even more so?

In this book you will discover the fundamental relationship between oxygen and the body. Improving fitness depends on enhancing the release of oxygen to your muscles, organs, and tissues. Increased oxygenation is not only healthy; it also enables greater exercise intensity with reduced breathlessness. In short, you will be able to discover better health and fitness as well as better performance.

If you do compete, you'll also enjoy your training and competition more than ever, because you'll be able to achieve more with less effort. Overall fitness and sports performance is usually limited by the lungs—not by the legs, the arms, or even the mind. As anyone who engages in regular exercise knows, the feeling of intense breathlessness during sporting activity dictates exercise intensity far more than muscle fatigue. The foundation of enjoying and improving physical exercise, therefore, is to ensure that breathing is optimally efficient.

Chronic Overbreathing

Scientific research, as well as the experience of thousands of people I have worked with, has shown me the vital importance of learning how to breathe correctly. The problem is that correct breathing, which should be everyone's birthright, has become extremely challenging in our modern society. We assume that the body reflexively knows how much air it needs at all times, but unfortunately this is not the case. Over the centuries we have altered our environment so dramatically that many of us have forgotten our innate way of breathing. The process of breathing has been warped by chronic stress, sedentary lifestyles, unhealthy diets, overheated homes, and lack of fitness. All of these contribute to poor breathing habits. These in turn contribute to lethargy, weight gain, sleeping problems, respiratory conditions, and heart disease.

Our ancestors lived on a natural diet in a far less competitive environment and physically worked hard, a lifestyle conducive to maintaining an efficient breathing pattern. Compare that to modern-day living, in which we spend hours slouched at a desk on computers and talking on phones, surviving on a rushed lunch of convenience food, trying to manage a seemingly neverending series of tasks and financial obligations.

Modern living gradually increases the amount of air we breathe, and while getting more oxygen into our lungs might seem like a good idea, it is in fact light breathing that is a testament to good health and fitness. Think of an overweight tourist and an Olympian both arriving for the Summer Games. As they picked up their luggage and carried it up a flight of stairs, whom would you expect to be huffing and puffing? Certainly not the Olympian.

The biggest obstacle to your health and fitness is a rarely identified problem: *chronic overbreathing*. We can breathe two to three times more air than required without knowing it. To help determine if you are overbreathing, see how many of these questions you answer "yes" to:

- Do you sometimes breathe through your mouth as you go about your daily activities?
- Do you breathe through your mouth during deep sleep? (If you are not sure, do you wake up with a dry mouth in the morning?)
- Do you snore or hold your breath during sleep?
- Can you visibly notice your breathing during rest? To find out, take a look at your breathing right now. Spend a minute observing the movements of your chest or abdomen as you take each breath. The more movement you see, the heavier you breathe.
- When you observe your breathing, do you see more movements from the chest than from the abdomen?
- Do you regularly sigh throughout the day? (While one sigh every now and again is not an issue, regular sighing is enough to maintain chronic overbreathing.)
- Do you sometimes hear your breathing during rest?
- Do you experience symptoms resulting from habitual overbreathing, such as nasal congestion, tightening of the airways, fatigue, dizziness, or light-headedness?

Answering yes to some or all of the questions above suggests a tendency to overbreathe. These traits are typical of what happens when the amount of air we breathe is greater than what we need. Just as we have an optimal quantity of water and food to consume each day, we also have an optimal quantity of air to breathe. And just as eating too much can be damaging to our health, so can overbreathing.

The unconscious habit of overbreathing has hit epidemic proportions all across the industrialized world, and it's highly detrimental to our health. Chronic overbreathing leads to loss of health, poor fitness, and compromised performance and also contributes to many ailments including anxiety, asthma, fatigue, insomnia, heart problems, and even obesity. It may seem strange that such a disparate range of complaints can be caused by or worsened by overbreathing, but the breath of life influences literally every aspect of our health.

The purpose of this book is to return you to how you were meant to live and breathe. I will teach you simple methods that will counteract bad breathing habits, unearthing a new well of cardiovascular fitness that will improve your overall health and well-being. Serious athletes will achieve new levels of performance, fitness enthusiasts will unleash untapped potential, and those who are still trying to manage their health will overcome barriers to a more healthful lifestyle.

But, as with all conditions, to arrive at the remedy it's crucial to first understand the ailment.

It is how you breathe during your daily life that determines how you breathe during physical exercise. Breathing too much air every minute, every hour, every day translates into excessive breathlessness during exercise. If our breathing is off during rest, it would be unreasonable to expect it to automatically correct itself during physical exercise. The seemingly innocuous tendency to breathe through the mouth during the day or night and breathe noticeably during rest means you will be more breathless during training and often limits your capacity to go faster and farther.

These poor breathing habits can be the difference between a healthy and vibrant life and an ill and feeble one. Overbreathing causes the narrowing of airways, limiting your body's ability to oxygenate, and the constriction of blood vessels, leading to reduced blood flow to the heart and other organs and muscles. These systemic impacts affect your health profoundly, whether you're a professional athlete or your main exercise is walking up the stairs of your house. Great sports careers can plateau or even be cut short by an athlete's overbreathing. The lungs let the individual down, and—no matter how strong the rest of the body is—unnecessary, excess breaths take their toll. As most athletes know, our lungs give out long before our arms and legs.

It all comes down to our need for that invisible yet vital basis for human life: oxygen. Here's the paradox: The amount of oxygen your muscles, organs, and tissues are able to use is not entirely dependent on the amount of oxygen in your blood. Our red blood cells are saturated with between 95 and 99 percent oxygen, and that's plenty for even the most strenuous exercise. (A few of my clients with serious pulmonary disease have a lower oxygen saturation level, but this is very rare.) What determines how much of this oxygen your body can use is actually the amount of carbon dioxide in your blood. You may remember from biology class that we breathe in oxygen and breathe out carbon dioxide, also called CO_2. Most people learn that carbon dioxide is just a waste gas that we exhale from our lungs, but it is not a waste gas. It is the key variable that allows the release of oxygen from the red blood cells to be metabolized by the body. This is called the Bohr Effect. Understanding and utilizing this physiological principle will allow you to stop overbreathing.

Discovered over a hundred years ago, the Bohr Effect explains the release of oxygen to working muscles and organs. Most people don't realize that the amount of carbon dioxide present in our blood cells determines how much oxygen we can use. The crux of it is this: How we breathe determines the levels of carbon dioxide present in our blood. When we breathe correctly, we have a sufficient amount of carbon dioxide, and our breathing is quiet, controlled, and rhythmic. If we are overbreathing, our breathing is heavy, more intense, and erratic, and we exhale too much carbon dioxide, leaving our body literally gasping for oxygen.

It's very intuitive: If we breathe better, increasing the amount of carbon dioxide inside us, then we can deliver more oxygen to our muscles and organs, including the heart and brain, and thus heighten our physical capacity. All we're really doing is assisting the body in working the way it was meant to work in the first place.

Bringing the Mountain to You

To understand how part two of the Oxygen Advantage works, let's look at an example most of us are familiar with: high-altitude training, a technique often used by elite athletes to improve their cardiovascular fitness and to improve their endurance. High-altitude training first came to the attention of coaches and athletes during the 1968 Summer Olympics, held in Mexico City at a height of 2,300 meters above sea level. Many competing athletes found that when they returned to sea level, their performance surpassed their previous personal best, prompting coaches to question whether athletes might perform better if they live or train at high altitude.

At high altitude the air is thin, which results in reduced atmospheric pressure of oxygen. The body adapts to this environment by increasing the number of red blood cells. Think of red blood cells as your very own Popeye's spinach, only they come from your body instead of out of a can. Upping the presence of red blood cells translates into improved oxygen delivery to the muscles, a reduction of lactic acid buildup, and stronger overall performance, including longer endurance and a lower risk of inflammation and injury. But of course the catch is that high-altitude training is not available to most of us—which brings me to the goal of this book.

You don't need to go to the mountain. The mountain can come to you.

I will show you how to make this happen through simple techniques that in effect take you up a mile high. By learning how to simulate high-altitude training, you will increase the oxygen-carrying capacity of your bloodstream, allowing your red blood cells to fuel new capabilities. Additionally, it will help you to sustain sharper psychological focus during physical activity as you become less conscious of the act of breathing. This will free you to devote more attention to maintaining proper form while exercising or formulating strategy in a competitive sport.

If you reduce your breathing and properly regulate the amount of air you take in, you will teach your body to breathe more efficiently, and you will become healthier. No matter what your athletic baseline is to begin with, better breathing will revolutionize your fitness, your endurance, and your performance. I know this for a fact because I have experienced it myself. I was once a chronic overbreather.

Back in 1997 I was an executive in the corporate world, but since childhood I had experienced poor health as a result of asthma. My identity was based on everything that I wasn't. I wasn't fit, I wasn't healthy, and I wasn't confident. I looked desperately for a solution to my health problems. Then I found it.

My life changed forever when I discovered the work of the late Dr. Konstantin Buteyko, a brilliant Russian physician who conducted groundbreaking research to determine the optimal breathing for astronauts during the Soviet Space Race. The Cold War had trapped his pioneering methods on the other side of the Iron Curtain, but, starting in the 1990s, they began to be known in the rest of the world. Using breathing exercises based on Buteyko's teachings, I successfully reversed my own sleep-disordered breathing and chronic asthma, fully recovering from the conditions I had suffered from my whole life. Inspired, I left my corporate job and trained directly under Dr. Buteyko. Thanks to his work, my life underwent a profound transformation. When you experience something like that, it's impossible not to want to share it. In my case, sharing it has become my passion and my profession.

Over the last thirteen years I have built on Dr. Buteyko's innovative approach and developed the Oxygen Advantage program not only for significantly improving asthma control but also for helping to improve anyone's health and fitness. I have worked with more than five thousand individuals, running the gamut from lifelong couch potatoes to ripped-ab Olympians.

I'd like to tell you three stories about people whose lives radically changed because they learned how to stop overbreathing. One is a competitive athlete, one is a newly converted fitness junkie, and one was just trying to lose weight and to become a little healthier.

Breathing Excessively

In the Croke Park arena in Dublin, where I'm from, more than eighty thousand fans regularly crowd into the stadium to watch their favorite football teams compete. Every game feels like Super Bowl Sunday at "the Croker." Irish football is more than a sporting event—it's a passion, a way of life, and a source of national pride. While the players are considered semiprofessional, management spends considerable sums of money on the latest sports technology for the team members and closely monitors their lifestyle and physiological parameters, 24-7. If a player eats a french fry late at night, management knows.

I met David when he was a rising star at Croke Park. He was twenty years old and trained five days a week with his team. He was in excellent physical shape, but he frequently succumbed to breathlessness and experienced nasal congestion and coughing. There was nothing more exhilarating for David than to play in front of a packed stadium, but after every game he was dogged by a barking cough, and his lungs felt like they were full of junk. He worked hard at his training and even harder at hiding his symptoms from his coaches and their electronic monitors. Finally David paid a visit to his doctor and was prescribed medication, which helped slightly, but he still struggled to keep up with his fellow players and still worried that he would be dropped from the team if his coaches found out about his difficulties.

When I first started working with David, he displayed all the characteristics of a person who was breathing far more than his body required. He breathed heavily and through his mouth, even while resting. He was getting oxygen into his lungs but too much, and he wasn't self-regulating in the way that was natural—and indispensable—for a competitive athlete. Through bad habits accumulated over the years, his body was out of sync with his breath, and he had become unable to meet his own needs for CO_2.

I went through my program with David, and he practiced the exercises exactly as described in this book: reduced breathing, breath holding during training, and keeping his mouth closed at night to train him to breathe through his nose. Today, David is one of the star players on his team and no longer has to hide his breathlessness from his coach. But he still has to hide his love of french fries.

Like David, many competitive athletes breathe excessively, regardless of how many years they have spent training. For some, no matter how hard they train, they will never attain their ideal fitness. In addition, in order to maintain their fitness, they will need to train more than their peers. The first time athletes hear of the effects of chronic overbreathing it can take a little while to sink in, although often it comes as a revelation, answering questions that have been at the backs of their minds for years and giving them a whole new understanding of their training. By incorporating simple practices into your existing training program, you can enjoy a more intense workout without putting extra strain on your lungs. One factor that differentiates elite athletes from others is their ability to exercise at a higher intensity with reduced breathlessness. This book will help you to understand the factors that allow oxygen to be released to organs and working muscles, enabling you to improve running economy (lowering of energy expended during running) and increase "VO_2 max" (the maximum capacity of the body to transport and use oxygen).

Throughout the years I have witnessed miraculous results with all types of athletes, including rugby players, soccer players, runners, cyclists, swimmers, and Olympic competitors. So many of these athletes suffered from excessive breathlessness, weak diaphragms, and inefficient breathing, and the difference efficient breathing brought to their athletic ability has been nothing short of amazing to witness. Developing body strength while ignoring breathing efficiency is counterproductive, and this book will show you how to build your respiratory stamina alongside any athletic training program.

Harnessing Explosive Athletic Potential

While David's story is powerful, don't think that better breathing techniques only benefit elite athletes. They can be just as transformative for "normal" people and, in fact, are often more so. Take the case of Doug.

Doug is a high-powered American professional in his midforties. Since childhood he battled asthma, and he never considered himself an athlete. Doug's brother, on the other hand, was the jock. When they were kids they would go to the park, where his brother played basketball with their dad while Doug just watched. Doug always felt like there was something wrong with his body. He did manage to row crew in college for a year, trying to follow in his father's footsteps, but after each training run, his lungs would scream for mercy. His aerobic capacity—or lack thereof—limited him, keeping an athletic lifestyle out of reach. But finally, when his father started to become frail, he resolved to take action so that he would be around for his own kids and grandkids.

Doug started running, but he fell into his habitual gasping for breath after just a handful of steps. He realized he needed to rebuild his cardiovascular fitness from the ground up, and that's when he contacted me. By incorporating the simple program described in this book into a busy work and family life, he started making progress. From being able to run only ten feet with his mouth closed, he progressed to running a 10K in a few months, then a half-marathon after a few more, and finally the Big Sur marathon less than a year after we began working together.

Doug needed to let go of lifelong breathing habits. Overbreathing distorted his self-understanding and turned him into someone who he wasn't. I needed to convey to Doug that inheriting the genetic predisposition toward asthma didn't mean he was resigned to a life of breathing problems. Asthma has been around for thousands of years, with records dating back as far as ancient Egypt. However, it has become much more prevalent since the 1980s, and considering that our gene pool doesn't change in forty years, it is necessary to look at lifestyle and address the impact that this has on our breathing. Currently almost one in ten adults and children have asthma, and if you add the number who have cyclist's cough, exercise-induced asthma, or other lung-limiting conditions, the number skyrockets.

Over the years, I have worked with thousands of people like Doug who have been diagnosed with asthma, and it's nearly always the same story: Explosive athletic potential is limited by a condition they don't think they can ever overcome. By not addressing the root problem, enthusiasts like Doug often devote their formidable willpower to training practices that inevitably lead them back to square one. It doesn't have to be this way. At first it may seem counterintuitive to think that implementing simple techniques over a short period of time can reverse decades of limitations, but that's how transformative correcting breathing is. With breath-holding exercises that unblock the nose and combat wheezing or coughing, nonprofessional athletes—even those with asthma—can lift their passions to an entirely new level.

You may not have major athletic goals. Some of us just want to get to a weight that makes us feel good when we look in the mirror. For many people struggling to attain this sense of satisfaction, the barrier is standing right there—not in front of them, but inside them, in the quantity of air they are taking in. Without correct breathing, it's like walking up a down escalator—you get nowhere.

Feeling Defeated

There wasn't a diet Donna hadn't tried. You've heard of them all: low-carb, South Beach, The Zone, Weight Watchers, Jenny Craig, Mediterranean, Atkins, Slim-Fast. You name it, she had been on it. Her medicine cabinet was filled with fat burners, carb blockers, and a host of appetite suppressants. For twenty-five years she had believed that with each new diet she would finally lose the extra forty pounds she carried, finally be able to step outside in something other than black, form-hiding clothing, and finally reclaim the health she enjoyed when she was younger. But after the initial enthusiasm of beginning a new diet faded, the weight she lost would return and a profound sense of failure would set in.

When Donna came to see me, she was defeated. She had spent thousands of dollars to slim down again and again, but she was still forty pounds overweight—and still miserable. She had tried just as many exercise programs as diets, but always ended up quitting because she ran out of breath after minimal exertion. As is the case for so many people, oxygen felt like an opponent rather than an ally. The sensation of intense breathlessness limited her physical endurance far more than muscle fatigue.

"I can't exercise because I'm too heavy," Donna said. "And I can't lose weight because I can't exercise." During the few times that Donna did visit the gym, she felt totally self-conscious and out of place. She panted on the treadmill while beautifully sculpted bodies decked out in form-fitting clothing jogged effortlessly on either side of her, adding another blow to her self-confidence.

It was a vicious cycle, but one I had seen many times before. Her body wasn't properly metabolizing oxygen. Donna needed a simple routine that wouldn't put excess stress on her body and breathing, but would give her fast, concrete results to keep her motivated and boost her confidence. I gave her simple breathing exercises and encouraged her to practice breathing through her nose while watching television or working at her desk.

In two weeks, Donna lost six pounds. She didn't modify her diet, but her breath-reduction exercises gave a kick-start to the oxygen levels in her blood, which caused her body to process foods more efficiently and naturally supressed her appetite. She benefited from one of the most startling aspects of my program: Substantial gains can be accomplished while literally sitting on the couch. Once you see this progress, however, the last thing you'll want to do is keep sitting.

Today Donna has lost thirty pounds, and, more important, she finds it much easier to keep them off. My work with her and so many other people in similar situations has nothing to do with what to eat or not eat. It's critical to take a step back from the table as well as the scale in order to get a big-picture view of the problem. Weight loss only occurs when the amount of calories we burn is greater than those we consume, and our breathing has a direct bearing on this process. By focusing on not just how much we eat but how much we breathe, we cause our consumption-to-burning ratio to even out. With properly oxygenated cells, our bodies operate more efficiently, even—or especially—in passive activities like sitting. A desire for more water and less processed food naturally follows. This is why diet work is not a part of this book. The only guidance I give to people like Donna is to eat when hungry and stop when satisfied, allowing self-control to come from the inside. Putting better breathing at the center of your health plan makes you look and feel better.

The Oxygen Advantage program detailed in this book is the culmination of my work with thousands of people like David, Doug, and Donna. It empowers people, regardless of their activity level, to improve their health, fitness, and performance—without training more or taking any drug or supplement. The program also provides readers with the ability to easily and accurately measure advances and ensures that exercise is done safely, reducing the risk of injury. Lastly, the Oxygen Advantage program can be tailored to anyone and any lifestyle, enabling you to seamlessly incorporate it into your daily obligations and exercise routine.

The chapters to come will provide knowledge and practical breathing techniques so that you can optimize oxygen release at a cellular level. The simple practices I will explain, while unknown to most athletes, have been utilized since ancient times to great effect. All you need to apply them is a straightforward understanding of how your breathing affects the oxygenation of your body

Part I of this book, The Secret of Breath, explains in greater detail the function of oxygen and CO_2 in your body and helps you to evaluate how fit you actually are. You'll learn about the importance of nose breathing over mouth breathing, as well as the first key technique that will begin to reverse overbreathing. I will also teach you about the ancient breathing secrets that have been used for centuries.

In Part II, The Secret of Fitness, you will learn about red blood cells and how utilizing them the way Olympic athletes do will carry you into a new realm of fitness. This section also introduces you to simulated high-altitude training and teaches you how to find "the zone" mentally as well as physically.

Part III, The Secret of Health, explores how better breathing naturally leads to weight loss and reduces the risk of sports-related injuries. It also explains the relationship between oxygenation and improved heart function. And for people prone to asthma, it gives you the tools to eliminate exercise-induced asthma.

Part IV takes everything you've learned and shows you how to construct your own personal Oxygen Advantage program. This section is geared toward specific groups based on their health and fitness. Breathing is generally an involuntary activity that we engage in unconsciously and rarely think about, but it's there always, every single moment we're alive, either helping us forward or holding us back. The point of this book is to elevate your awareness of how you can harness your breath to reclaim your body's natural ability to breathe in a way that will help you to achieve lifelong health and fitness, whether you are running to catch up with your kids or running to win a gold medal. My promise is that by applying the concepts and simple exercises in this book, each and every person, whether they consider themselves an athlete or not, will be able to attain tangible and profound improvements to their health, fitness, and performance within just a few weeks. Isn't it time you did more—conditioning, winning, living—with less effort?

PART I

The Secret of Breath

CHAPTER 1
The Oxygen Paradox

Sport has always been the great love of Don Gordon's life. He loved everything about it—the sweat, the competition, the adversity, the triumph. Growing up, he attended many races and football games with his father, watching his favorite competitors and aspiring to be just like them. Nothing compared to the atmosphere of a good game: the excitement of the fans, the shouts of encouragement (or profanities, according to the progress of the game), and always the belief and the hope that one day he would be just like the athletes he idolized.

As a teenager, cycling was Don's sport. He spent hours training on his bike, but he could never quite keep up with his fellow cyclists. He tired quickly and more often than not found himself breathless, watching from a distance as his friends rode their bikes farther and longer than he ever could. As time passed Don reluctantly gave up his dream of competing like the athletes he so admired as a boy. He finally accepted that there was no place for him in the world of competitive cycling.

Twenty years later Don had become director of European operations in a leading American technology firm. While on a trip to Europe, he happened across my Oxygen Advantage program. Because Don had tried so many things before, he was skeptical, but decided to give it a shot. He got in touch with me, and in our first session together I gave him the crash course I gave you in the introduction to this book. He had never considered the relationship between physical capacity and correct breathing, but with a new understanding of the potential of improved body oxygenation, Don began practicing the exercises I gave him. Within days he felt better and had greater energy than ever before. Flash-forward to today: Don has been free of wheezing, allergies, and medications for more than seven years. He is also now a competitive long-distance cycler, and in his most recent race he finished first in his age division. And here's the best part: At the age of fifty-eight, he had the twenty-ninth-fastest overall time across a wide field of 320 competitors, including exceedingly fit twenty- to thirty-year-olds. He has finally come to resemble the athletes he so looked up to as a boy.

Getting his breathing right was the key to changing everything for Don. Breathing is natural and involuntary. We don't have to remember to breathe in and out. If we did, either we would have to devote all of our time and energy to it, or we would have stopped living a long time ago. But while breathing is our most instinctive act, many factors of modern life negatively affect our breathing—and worse still, we're highly misinformed about how our breathing affects our bodies during physical exertion. During a presentation to a group of runners who were due to compete in the Dublin city marathon the next day, I posed this question: "Who here believes that taking a large breath into the lungs during rest will increase oxygen content of the blood?" Without hesitation, 95 percent of the runners raised their hands. They were wrong, but they aren't alone—this belief is widespread in the world of sports and fitness. But taking a large breath into the lungs during rest will *not* increase oxygen content. It is exactly the wrong thing to do if you seek greater endurance.

Based on this misconception, many athletes adopt the practice of intentionally taking deep breaths during rest and training, and especially when their bodies are overtaxed. By doing so, however, they in fact limit and sometimes even diminish their performance.

As I will explain, however, it is possible to reverse these negative influences of modern life and to condition our body to breathe healthy amounts of air during resting periods. By doing so, we ensure that the right amount of oxygen is powering our muscles, lungs, and heart. This will lead to reduced breathlessness during physical exercise, which in turn will make improved fitness more attainable. Better breathing is the gateway to a new realm of health.

Breathe Right to Maximize Oxygenation of Your Brain, Heart, and Other Working Muscles

Before you start the Oxygen Advantage exercises, it is important for you to have a basic understanding of the respiratory system and the role of carbon dioxide in your body. If you would like to bypass the science, you can go straight to chapter 2, but the more you know, the more you can work with your body and not against it.

The Respiratory System

Your respiratory system comprises the parts of your body that deliver oxygen from the atmosphere to your cells and tissues and transport the carbon dioxide produced in your tissues back into the atmosphere.

Your respiratory system contains everything you need to adequately oxygenate your body for exercise and high-performance sports . . . so long as you allow it to function properly. When we breathe, air enters the body and flows down the windpipe (trachea), which then divides into two branches called *bronchi:* One branch leads to the right lung, the other to the left. Within your lungs, the bronchi further subdivide into smaller branches called *bronchioles,* and eventually into a multitude of small air sacs called *alveoli.* To visualize this complex system, imagine an upside-down tree. Your trachea is the trunk, and the bronchi form two large branches at the top of it, from which the smaller branches of the bronchioles grow. At the end of these branches are the "leaves"—the round sacs of the alveoli, which transport oxygen into the blood. It is quite a striking example of evolutionary balance and beauty that the trees around us that give off oxygen and the trees in our lungs that absorb it share a similar structure.

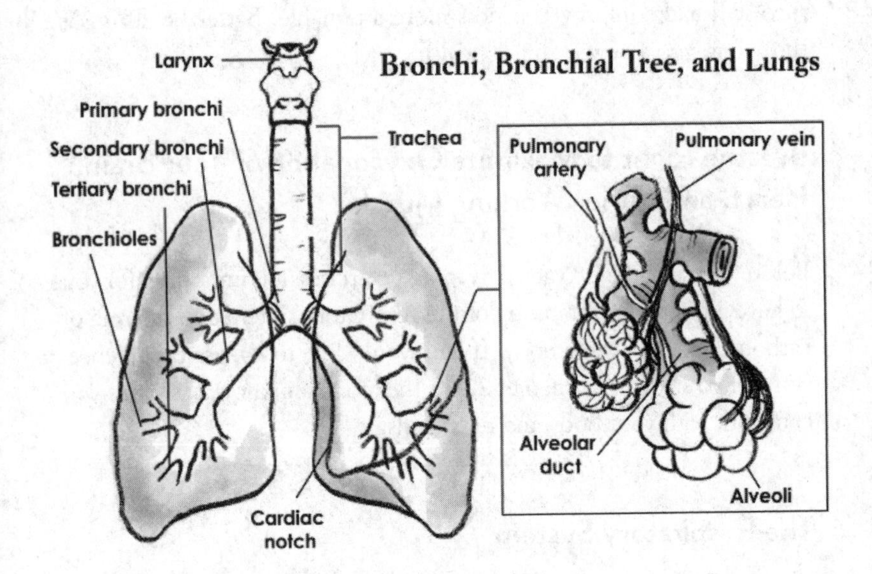

The lungs contain approximately 300 million alveoli, each of which is surrounded by tiny blood vessels called *capillaries*. To put this immense number in context, the area of contact between your alveoli and blood capillaries is equivalent to the size of a tennis court. This large, impressively contained surface provides the potential for an extremely efficient transfer of oxygen to the blood.

As I've explained, oxygen is the fuel that muscles need to work efficiently. It is, however, a common misconception that breathing in a larger volume of air increases the oxygenation of the blood. It is physiologically impossible to increase the oxygen saturation of the blood in this way, because the blood is almost always already fully saturated. It would be like pouring more water into a glass that is already filled to the brim. But what is oxygen saturation exactly, and how does it relate to properly oxygenating our muscles?

Oxygen saturation (SpO_2) is the percentage of oxygen-carrying red blood cells (hemoglobin molecules) containing oxygen within the blood. During periods of rest the standard breathing volume for a healthy person is between 4 and 6 liters of air per minute, which results in almost complete oxygen saturation of 95 to 99 percent. Because oxygen is continually diffusing from the blood into the cells, 100 percent saturation is not always feasible. An oxygen saturation of 100 percent would suggest that the bond between red blood cells and oxygen molecules is too strong, reducing the blood cells' ability to deliver oxygen to muscles, organs, and tissues. We need the blood to release oxygen, not hold on to it. The human body actually carries a surplus of oxygen in the blood—75 percent is exhaled during rest and as much as 25 percent is exhaled during physical exercise. Increasing oxygen saturation to 100 percent has no added benefits.

The idea of taking bigger breaths to take in more oxygen is akin to telling an individual who is already eating enough food to provide their daily caloric needs that they need to eat more. Many of my students initially have a hard time grasping this. For years they have been indoctrinated with the "benefits" of taking deep breaths by well-meaning stress counselors, yoga practitioners, physiotherapists, and sports coaches, not to mention the Western media. And it's easy to see why this belief is perpetuated: Taking a large breath can actually feel good, even if it can actually be bad for you. Just as a cat enjoys a good stretch following a midday nap, taking a big breath into the lungs stretches the upper part of the body, allowing a feeling of relaxation to follow. But this leads many to believe that with breathing, bigger is better.

Regulation of Breathing

There are two main aspects to the way you breathe: the *rate* or number of breaths you take in the space of 1 minute and the *volume* or amount of air drawn into your lungs with each breath. Although the two are separate, one generally influences the other.

The volume of each breath of air we inhale and exhale is measured in liters, and measurements are usually taken over 1 minute.

In conventional medicine the accepted number of breaths a healthy person takes during that minute is 10 to 12, with each breath drawing in a volume of 500 milliliters of air, for a total volume of 5 to 6 liters. To visualize this amount of air, imagine how much air would be contained in about three empty 2-liter soft drink bottles. If a person is breathing at a higher rate—at 20 breaths per minute, for example—then the volume will also be higher. But overbreathing doesn't just come from an elevated rate. A lower rate can have the same effect if the individual is taking in too much air with each breath; 10 large breaths of 1,000 milliliters would also be evidence of overbreathing. In the next chapter, you will be able to measure your own relative breathing volume using a breath-hold test called the Body Oxygen Level Test, or BOLT.

So how do we ensure that we breathe correctly so as to make optimal use of our amazing respiratory system? As odd as this may seem, it's not oxygen that exerts the primary influence on your breathing efficiency, but carbon dioxide.

The rate and volume of breathing is determined by receptors in the brain that work in a way similar to a thermostat regulating the heating system in a home. However, instead of monitoring fluctuations in temperature, these receptors monitor the concentration of carbon dioxide and oxygen in your blood, along with the acidity or pH level. When levels of carbon dioxide increase above a certain amount, these sensitive receptors stimulate breathing in order to get rid of the excess gas. In other words, the primary stimulus to breathe is to eliminate excess carbon dioxide from the body.

Carbon dioxide is an end product of the natural process of breaking down the fats and carbohydrates we eat. CO_2 is returned from the tissues and cells to the lungs via blood vessels, and any excess is exhaled. Crucially, however, part of your body's quotient of carbon dioxide is retained when you exhale. Correct breathing both relies on and results in the right amount of carbon dioxide being retained in your lungs. Understanding this is just as important for serious athletes as it is for anyone interested in basic fitness or in weight management.

Think of it this way: CO_2 is the doorway that lets oxygen reach our muscles. If the door is only partially open, only some of the oxygen at our disposal passes through, and we find ourselves gasping during exercise, often with our limbs cramping. If, on the other hand, the door is wide open, oxygen flows through the doorway and we can sustain physical activity longer and at a higher intensity. But to understand how our breathing works we must dig a bit deeper into the crucial role carbon dioxide plays in making it as efficient as possible.

Chronic hyperventilation or overbreathing simply means the habit of breathing a volume of air greater than that which your body requires. It does not necessarily manifest as dramatic symptoms, such as the panting a person might experience during a panic attack. When we breathe in excess of what we require, too much carbon dioxide is exhaled from the lungs and, hence, is removed from the blood. It forces that door to a more closed position, making it harder for oxygen to pass through. Breathing too much for short periods of time is not a significant problem, as no permanent change in the body occurs. However, when we breathe too much over an extended period of days to weeks, a biochemical change takes place inside us that results in an increased sensitivity or lower tolerance to carbon dioxide. With this lower set point, breathing volume remains above normal as the receptors in the brain continuously stimulate breathing in order to get rid of carbon dioxide that is perceived to be in excess of the receptor's programmed limits. The result is the habit of chronic overbreathing or chronic hyperventilation, with all its negative manifestations. In other words, certain circumstances can train our body to breathe in such a way that goes against its own interests. To counteract these bad habits, you must retrain yourself to breathe better.

I often ask my groups of students this question: "Who feels that they are more tired than they should be?" Usually about 80 percent raise their hands. My job is to help them understand why. With the aid of a pulse oximeter, I have measured the oxygen saturation of thousands of people, and the vast majority display normal blood oxygen saturations of between 95 and 99 percent.* Why would this be? Their blood oxygen saturations are normal, yet they constantly feel tired. The problem is not a lack of oxygen in the blood, but that not enough oxygen is being released from the blood to tissues and organs, including the brain, resulting in feelings of lethargy and exhaustion. This happens because too much carbon dioxide has been expelled from the body. As we shall see further on, habitual overbreathing influences the release of oxygen from red blood cells, the consequences of which can affect day-to-day well-being as well as performance during exercise. This ties back to the Bohr Effect, which I touched on in the introduction and will expand on in a few pages.

One's breathing volume can be two or three times the required amount without it being overtly noticeable. Once the pattern of overbreathing is established, it is often maintained by an occasional deep breath or sigh. When such a habit becomes ingrained both mentally and physically, you will breathe in excess of what is required every minute, every hour, and every day. This subtle alteration to your body's natural functioning can hinder you greatly. And it doesn't just happen while we're conscious; many people sleep with their mouth open, and whether they realize it or not, this drags down their physical and mental energy.

So why is it that the benefits of light breathing are relatively unknown? It is difficult to know the exact answer, although a number of points are worth bearing in mind. The first is that air is weightless and therefore difficult to measure, and breathing can change quickly and effortlessly during the measuring process. The second is that doctors learn how oxygen is released from the red blood cells early on in their studies—the Bohr Effect is described in most basic medical school physiology textbooks—so it is possible that this information is simply forgotten by the time of graduation. Another reason may be that overbreathing affects each person individually, resulting in a wide variety of problems that may not necessarily appear to be connected, from car-

* From time to time, I do see an individual with lower blood oxygen saturation, but this generally results from severe lung obstruction such as chronic obstructive pulmonary disease.

diovascular, respiratory, and gastrointestinal issues to general exhaustion. To add even more confusion, not everyone who overbreathes will develop obvious symptoms, as the effects of hyperventilation depend on genetic predisposition.

Finally, given the lack of awareness of the relationship between breathing volume and health, so many chronic overbreathers have learned to tolerate the stunted levels of energy and fitness incorrect breathing leaves them with in day-to-day life. But shaking ourselves out of this complacent attitude toward our breath and putting it at the center of our health often produces more dramatic changes than any diet.

So how can we regulate the amount of air we breathe in order to optimize our fitness and athletic performance? As you know by now, the vital ingredient is carbon dioxide.

Carbon Dioxide: Not Just a Waste of Gas

The concentration of carbon dioxide in the earth's atmosphere is very low, which means that we don't carry it into our lungs when we breathe. Instead we produce it in tissue cells during the process of converting food and oxygen into energy. Maintaining a correct breathing volume ensures that the ideal amount of carbon dioxide remains in the lungs, blood, tissues, and cells.

Carbon dioxide performs a number of vital functions in the human body, including:

- Offloading of oxygen from the blood to be used by the cells.
- The dilation of the smooth muscle in the walls of the airways and blood vessels
- The regulation of blood pH.

Delivery of Oxygen from the Blood to the Muscles and Organs

Hemoglobin is a protein found in the blood, and one of its functions is to carry oxygen from the lungs to the tissues and cells. A fundamental element of the Oxygen Advantage technique is to understand the Bohr Effect—the way in which oxygen is released from hemoglobin and delivered to the muscles and organs. This process forms the core of unlocking your body's true fitness potential, allowing you to raise your game and achieve the results you really want.

The Bohr Effect was discovered in 1904 by the Danish physiologist Christian Bohr (father of Niels Bohr, the Nobel Prize–winning physicist—and footballer). In the words of Christian Bohr, "The carbon dioxide pressure of the blood is to be regarded as an important factor in the inner respiratory metabolism. If one uses carbon dioxide in appropriate amounts, the oxygen that was taken up can be used more effectively throughout the body."

The crucial point to remember is that hemoglobin releases oxygen *when in the presence of carbon dioxide.* When we overbreathe, too much carbon dioxide is washed from the lungs, blood, tissues, and cells. This condition is called *hypocapnia,* causing the hemoglobin to hold on to oxygen, resulting in reduced oxygen release and therefore reduced oxygen delivery to tissues and organs. With less oxygen delivered to the muscles, they cannot work as effectively as we might like them to. As counterintuitive as it may seem, the urge to take bigger, deeper breaths when we hit the wall during exercise does not provide the muscles with more oxygen but effectively reduces oxygenation even further. In contrast, when breathing volume remains nearer to correct levels, the pressure of carbon dioxide in the blood is higher, loosening the bond between hemoglobin and oxygen and facilitating the delivery of oxygen to the muscles and organs. John West, author of *Respiratory Physiology,* tells us that "an exercising muscle is hot and generates carbon dioxide, and it benefits from increased unloading of O_2 [oxygen] from its capillaries." The better we can fuel our muscles with oxygen during activity, the longer and harder they can work. In light of the Bohr Effect, overbreathing limits the release of oxygen from the blood, and in turn affects how well our muscles are able to work.

Dilation and Constriction of Airways and Blood Vessels

Breathing too much can also cause reduced blood flow. For the vast majority of people, 2 minutes of heavy breathing is enough to reduce blood circulation throughout the body, including the brain, which can cause a feeling of dizziness and light-headedness. In general, blood flow to the brain reduces proportionately to each reduction in carbon dioxide. A study by Dr. Daniel M. Gibbs, which was published in the *American Journal of Psychiatry* to assess arterial constriction induced by excessive breathing, found that the diameter of blood vessels reduced in some individuals by as much as 50 percent. Based on the formula πr^2, which measures the area of a circle, blood flow decreases by a factor of four. This shows you how radically overbreathing can affect your blood flow.

Most people will have experienced constriction of blood flow to the brain resulting from overbreathing. It doesn't take very long to feel the onset of dizziness from taking a few big breaths in and out through the mouth. Similarly, many individuals who sleep with their mouths open may find it difficult to get going in the morning. However long they sleep, they are still tired and groggy for the first few hours after waking. It is well documented that habitual mouth breathing during waking and sleeping hours results in fatigue, poor concentration, reduced productivity, and a bad mood. Hardly an ideal recipe for quality living or a productive exercise program.

The same can also be true of individuals whose occupation involves considerable talking, such as schoolteachers or salespeople. People in these professions are often all too aware of how tired they feel after a day of work, but the exhaustion that follows endless business meetings is not necessarily due to mental or physical effort—more likely it is a result of the effects of elevated breathing levels during excessive talking. It is normal for breathing to increase during physical exercise as the body demands more oxygen to convert food into energy. However, in the case of talking, breathing increases without an actual need for more oxygen, causing a disturbance to blood gases and reducing blood flow.

Depending on genetic predisposition to asthma, the loss of carbon dioxide in the blood can also cause the smooth muscles of the airways to constrict, resulting in wheezing and breathlessness. However, an increase of carbon dioxide opens up the airways to allow a better oxygen transfer to take place and has been shown to improve breathing for persons diagnosed with asthma. But at the end of the day, we're all operating on the same spectrum, with good breathing at one end and bad breathing on the other. It's not just people with asthma who benefit from less constricted airways. The feeling of chest tightness, excessive breathlessness, cough, and the inability to take a satisfying breath is experienced by many athletes, including those without a prior history of asthma, but can be eliminated by simply improving the way you breathe.

The Regulation of Blood pH

In addition to determining how much oxygen is released into your tissues and cells, carbon dioxide also plays a central role in regulating the pH of the bloodstream: how acidic or alkaline your blood is. Normal pH in the blood is 7.365, and this level must remain within a tightly defined range or the body is forced to compensate. For example, when the blood's pH becomes more alkaline, breathing reduces to allow carbon dioxide levels to rise and restore pH. Conversely, if the pH of the blood is too acidic (as it is when you overconsume processed foods), breathing increases in order to offload carbon dioxide as acid, allowing pH to normalize. Maintaining normal blood pH is vital to our survival. If pH is too acidic and drops below 6.8, or too alkaline and rises above 7.8, the result can be fatal. This is because pH levels directly affect the ability of our internal organs and metabolism to function.

The pH–CO_2 Link

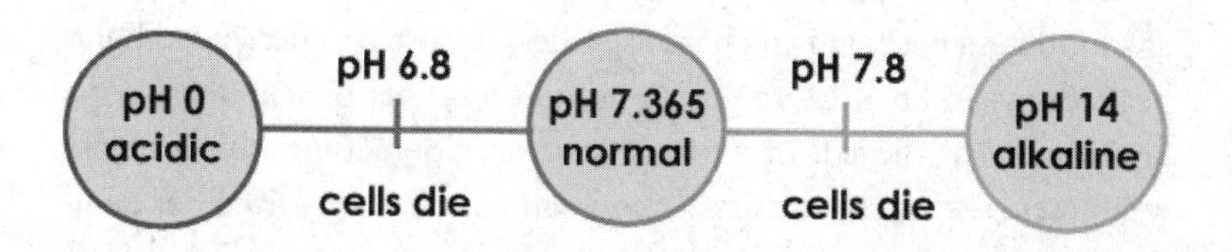

Scientific evidence clearly shows that carbon dioxide is an essential element not just in regulating our breathing, optimizing blood flow, and releasing oxygen to the muscles, but also maintaining correct pH levels. In short, our body's relationship with carbon dioxide determines how healthy we can be, affecting nearly every aspect of how our body functions. Better breathing allows carbon dioxide to ensure that all the interlocking parts of our system work together in harmony, allowing us to achieve our maximum potential in sporting performance, endurance, and strength.

Without the requisite amount of CO_2 in the blood, blood vessels constrict and hemoglobin cannot release oxygen into the tissues; without the requisite amount of oxygen, working muscles do not perform as effectively as they should. We become breathless, or hit a wall in our capabilities. It becomes a cycle: It's not just the breathless exertion that leads to panting. It's the panting that leads to breathless exertion. In the chapters to come, you'll learn how to break this cycle and build a new, positive one.

Eliminating overbreathing is the key to harnessing the potential of the CO_2 you already have inside you. Knowing how your respiratory system works is the first step in this empowering process, as it was for Alison, an amateur athlete who was an avid cyclist.

I met Alison when she was thirty-seven years old, and she had been cycling seriously since her late teens. She trained two to three times each week without fail, cycling up to 37 miles during each session. Cycling allowed Alison to have her own time, to leave her thoughts and worries behind, and to get out into nature and feel the breeze on her face.

Despite her years of regular training, Alison was experiencing excessive breathlessness and a desperate need for air even while cycling at a moderate pace. During her long rides, she often experienced lightheadedness and nausea, requiring her to get off her bike and wait for a few minutes by the side of the road to recover. Sometimes this problem was so severe that she felt like throwing up or fainting. Given her dedication to her training, she didn't understand why she wasn't in better condition, like her more fit cycling companions.

These bouts of nauseous dizziness continued, so Alison visited her doctor and then a specialist. They both ruled out asthma and any heart problems, giving her a clean bill of health. But the problem didn't resolve itself, and Alison's anxiety grew. She knew something was wrong, even though the medical tests and examinations revealed nothing.

A local sports coach put Alison in touch with me, and I immediately recognized signs of habitual mouth breathing, including excessive breathing movements from the upper chest. She sighed regularly and often felt short of air. Her bad breathing occurred not just while exercising, but in her everyday life, so she had created a self-reinforcing cycle that severely limited her abilities. While most health professionals would not give a moment's thought to Alison's breath, I was in no doubt that better breathing was the answer to her problem.

Alison was enormously relieved when I made her aware of her overbreathing habit as the root of her symptoms. She understood straightaway that if she was breathing too much during her everyday activities, then it stood to reason that her breathing would increase proportionately during sports, leading to excessive breathlessness. As is the case with many people, and not just athletes like Alison, overbreathing had thrown off her whole system. By losing the carbon dioxide her body so badly needed in order to send oxygen to her heart, other muscles, lungs, and head, she had hamstrung her own abilities. The breathlessness caused by mouth breathing had created a vicious cycle in which Alison felt the need to take in larger breaths to cope, resulting in a breathing volume that was even further increased.

Following two weeks of practicing the various exercises in this book, Alison reduced her breathlessness, and her nausea and fainting stopped. Her fitness levels and health also improved remarkably; she felt calmer, slept better, and had more energy throughout the day. Of course, not everyone with overbreathing experiences blackouts, as the effects of overbreathing will depend on genetic predisposition (which we'll discuss in chapter 13), but in all cases there will be some negative symptoms to be found—often unexplainable by doctors and specialists, as was the case with Alison. As the late chest physician Claude Lum explained, breathing too much "presents a collection of bizarre and often apparently unrelated symptoms, which may affect any part of the body, and any organ or any system." It's crucial to identify overbreathing as soon as possible, so as not to arrive at the extreme symptons Alison found herself battling.

In the next chapter we will look at a very simple way to measure our tolerance to carbon dioxide and relative breathing volume, and what this means to our health and sports performance. Finally, and more important, we will begin to learn the first steps to improving body oxygenation.

CHAPTER 2

How Fit Are You Really? The Body Oxygen Level Test (BOLT)

If you took a moderately paced run alongside an elite athlete, you would expect that his breathing would be light, rhythmic, easy, and effortless. You certainly wouldn't expect him to be huffing and puffing like a steam train. In fact, studies have shown that athletes experience up to 60 percent less breathlessness than untrained people when performing the same amount of exercise.

It is this feeling of intense breathlessness during exercise that often limits our ability to go faster and farther, and lighter breathing can be extremely advantageous to improving your performance. Being able to perform physical exercise with easy, slow breathing is not only a mark of good fitness; it is also healthier and safer.

During strenuous physical exercise, the consumption of oxygen increases, leading to a slightly reduced concentration of O_2 in the blood. At the same time, increased muscle activity and metabolic rate produces more carbon dioxide, causing an increased concentration of CO_2 in the blood.

As we have already explored, each breath we take is influenced by the continuing pressure of carbon dioxide (and to a lesser extent by the pressure of oxygen) within arterial blood vessels. When there is an increase of carbon dioxide and a decrease of oxygen, breathing is stimulated.

A very simple way to experience the effect carbon dioxide has on our stimulus to breathe is as follows: Gently exhale through your nose and pinch your nose with your fingers to hold your breath. As you hold your breath, carbon dioxide accumulates in the blood, and after a short while the receptors in the brain and neck signal the breathing muscles to resume breathing to get rid of the excess. You will begin to feel these signals in the form of contractions of the muscles in your neck and stomach, along with an urge to take in air. Let go of your nose and resume breathing through the nose when you feel the first messages of your body telling you to take a breath. It's important to bear in mind at this point that the purpose of breathing is to get rid of the excess carbon dioxide, and not to get rid of as much as possible. Overbreathing for a period of days and weeks, however, removes more carbon dioxide than is necessary, increasing the sensitivity of the brain's receptors.

The sensitivity of your receptors to carbon dioxide and oxygen will have implications for the way your body copes with physical exercise. When your breathing receptors have a strong response to carbon dioxide and reduced pressure of oxygen in the blood, your breathing will be intense and heavy. Your body will have to work much harder to maintain this increased breathing volume, but because overbreathing causes carbon dioxide levels to drop, less oxygen will be delivered to working muscles. The result? Overexertion, a disappointing performance, and possible injury.

Conversely, having a greater tolerance to carbon dioxide not only reduces breathlessness but also allows for much more effective delivery of oxygen to your working muscles during exercise. When breathing receptors are less sensitive to carbon dioxide levels, you will experience a reduction in breathlessness as your body is able to work harder with far less effort; breathing will be lighter during both rest and physical exercise. Efficient breathing means that fewer free radicals are produced, reducing the risk of inflammation, tissue damage, and injury.

Free radicals (or oxidants) are formed when the oxygen we breathe is converted into energy. During exercise, breathing markedly increases, resulting in an increased production of free radicals. Free radicals are a part of normal bodily function and are only of concern when there is an imbalance between these oxidants and the antioxidants that neutralize them. Oxidative stress occurs when there are too many free radicals in your system; left unchecked by antioxidants, free radicals attack other cells, causing inflammation, muscular fatigue, and overtraining.

It has been said that one of the main differences between endurance athletes and nonathletes is their response to low pressures of oxygen (hypoxia) and higher levels of carbon dioxide (hypercapnia). In other words, endurance athletes are able to tolerate a greater concentration of carbon dioxide and lower concentration of oxygen in the blood during exercise. Intense physical exercise results in increased consumption of oxygen and increased production of carbon dioxide, so it is vitally important that athletes are able to cope well with changes to these gases.

In order to attain outstanding performance during sports, it is essential that your breathing does not react too strongly to increased concentrations of carbon dioxide and decreased concentrations of oxygen. Over time, intense physical training will help to condition the body to better tolerate these changes, but a more effective method can be found in the pages of this book. The breathing exercises outlined in the Oxygen Advantage program can easily be incorporated into any form of exercise, no matter what your fitness level and even if you are laid up with an injury. You can even improve your fitness using a simple 10-minute exercise while sitting down.

Increasing VO_2 Max

A performance-related term you need to know is maximal oxygen uptake, or VO_2 max. This simply refers to the maximum capacity of your body to transport and utilize oxygen in 1 minute during maximal or exhaustive exercise. VO_2 max is one factor that can determine an athlete's capacity to sustain physical exercise and is considered to be the best indicator of cardiorespiratory endurance and aerobic fitness. In sports that require exceptional endurance, such as cycling, rowing, swimming, and running, world-class athletes typically have a high VO_2 max. Furthermore, the goal of most endurance programs is to increase VO_2 max.

Studies have shown that athletic ability to perform during increased carbon dioxide and reduced oxygen pressure corresponds to maximal oxygen uptake. In other words, the ability to tolerate higher concentrations of carbon dioxide in the blood means a higher VO_2 max can be achieved, culminating in better delivery and utilization of oxygen by the working muscles.

Without a doubt, regular physical training, correctly applied, helps to reduce the body's response to carbon dioxide, enabling greater exercise intensity and improving VO_2 max. During physical exercise, increased metabolic activity produces higher levels of carbon dioxide than normal. Over time, these raised CO_2 levels condition the breathing receptors, which results in easier, lighter breathing during exercise and better oxygenation of the muscles. Being able to deliver oxygen more efficiently during high-intensity exercise leads to a higher VO_2 max—something most athletes strive toward during their regular training.

In chapter 7 you will learn how to simulate high-altitude training. When the breath is held, oxygen saturation in the blood decreases, leading to increased production of red blood cells to offset the drop. Since red blood cells carry oxygen, having a greater quantity in your blood will also lead to an increase in aerobic capacity and VO_2 max. Along with VO_2 max, another performance measurement that is highly regarded by athletic coaches is running economy. This is defined by the amount of energy or oxygen consumed while running at a speed that is less than maximum pace. Typically, the less energy required to run at a given pace, the better—if your body is able to use oxygen efficiently, it is indicative of a high running economy.

There is a strong association between running economy and distance running performance in elite runners, where running economy is regarded as a better predictor of performance than VO_2 max. For this reason, sports scientists, coaches, and athletes are keen to apply techniques that can improve running economy, such as strength training and high-altitude training. However, a third and far more widely accessible method of boosting running economy is to practice breath-hold techniques, which have been proven to improve respiratory muscle strength and endurance. Researchers investigating reduced breathing found that running economy could be improved by a remarkable 6 percent following a brief course of breath-hold training.

At this point you might be thinking that if physical training conditions the body to tolerate higher concentrations of carbon dioxide and lower oxygen, then why go to the bother of practicing this program? Good question, but in our modern world, it is literally impossible to isolate ourselves from the factors that negatively affect breathing, and even many highly conditioned athletes breathe too heavily during rest; their inhalations and exhalations are noticeable and often from the upper chest. They may perform brilliantly, but they could be performing even better.

It would be disingenuous to expect breathing to be efficient during sport if breathing during times of rest is inefficient. If your breathing is wrong during the normal course of your day, how can it be right during the hour or two spent doing exercise? It just does not work that way. The Oxygen Advantage program focuses on retraining us in the way we breathe during rest and low-intensity exercise as well as moderate to high-intensity activity. This method helps to create good habits and brings lifelong benefits to your breathing, no matter what your level of fitness or your preferred sport.

In the next section you can determine your sensitivity to carbon dioxide using the Body Oxygen Level Test (BOLT), which measures the length of a comfortable breath hold. First you will learn your current condition, then you'll learn how the Oxygen Advantage can help you improve your sleep, concentration, and energy levels; attain a calmer disposition; reduce breathlessness during physical exertion; and increase that ever-coveted VO_2 max.

The Body Oxygen Level Test (BOLT)

As far back as 1975, researchers noted that the length of time of a comfortable breath hold served as a simple test to determine relative breathing volume during rest and breathlessness during physical exercise. The Body Oxygen Level Test (BOLT) is a very useful and accurate tool for determining this relative breathing volume. BOLT is simple, safe, involves no sophisticated equipment, and can be applied at any time. BOLT differs from other breath-hold tests because it represents the length of time until the first definite desire to breathe. Holding the breath until you feel the first natural desire to breathe provides useful information on how soon the first sensations of breathlessness take place and is a very useful tool for the evaluation of breathlessness. Other breath-hold tests tend to focus on the maximum time you can hold your breath, but this measurement is not objective as it can be influenced by willpower and determination.

Athletes possess bucketloads of willpower and determination, so there is no doubt that many of us will be tempted to measure our BOLT score by holding the breath for as long as possible. But if you are serious about improving your breathing efficiency and VO_2 max using the breath-hold exercises in this book, I urge you to follow the instructions carefully and measure your BOLT correctly—by holding your breath only until the first distinct urge to breathe is felt.

In short, the lower the BOLT score, the greater the breathing volume, and the greater your breathing volume, the more breathlessness you will experience during exercise.

To obtain an accurate measurement, it's best to rest for 10 minutes before measuring your BOLT score. Read the instructions carefully first and have a timer on hand. You can measure your BOLT now:

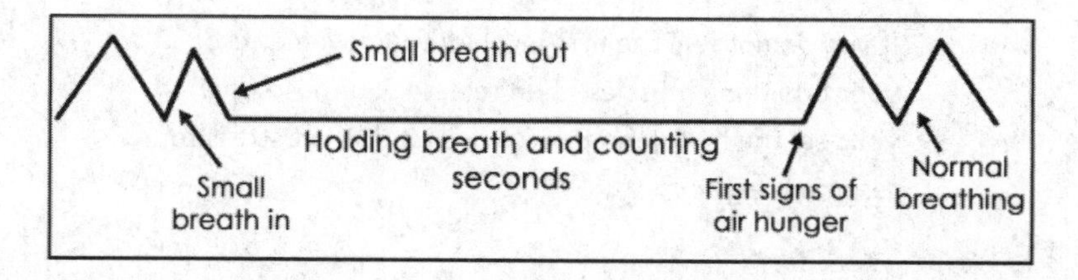

1. Take a normal breath in through your nose and allow a normal breath out through your nose.
2. Hold your nose with your fingers to prevent air from entering your lungs.
3. Time the number of seconds until you feel the first definite desire to breathe, or the first stresses of your body urging you to breathe. These sensations may include the need to swallow or a constriction of the airways. You may also feel the first involuntary contractions of your breathing muscles in your abdomen or throat as the body gives the message to resume breathing. (*Note that BOLT is not a measurement of how long you can hold your breath but simply the time it takes for your body to react to a lack of air.*)
4. Release your nose, stop the timer, and breathe in through your nose. Your inhalation at the end of the breath hold should be calm.
5. Resume normal breathing.

Please be aware of the following important points when measuring your BOLT score:

- The breath is taken after a gentle exhalation.
- The breath is held until the breathing muscles first begin to move. You are not measuring the maximum time that you can hold your breath.
- If you do not feel the first involuntary movements of your breathing muscles, then release your nose when you feel the first definite urge or first distinct stress to resume breathing.

- The BOLT is not an exercise to correct your breathing.
- Remember that measuring your BOLT involves holding your breath only until you feel the first involuntary movements of your breathing muscles. If you need to take a big breath at the end of the breath hold, then you have held your breath for too long.

How the Body Oxygen Level Test (BOLT) Works

When you hold your breath, you prevent oxygen from entering your lungs and prevent carbon dioxide from being expelled into the atmosphere. As the breath hold continues, carbon dioxide accumulates in the lungs and blood while oxygen levels slightly decrease. Since carbon dioxide is the primary stimulus for breathing, the length of your breath-hold time is influenced by how much carbon dioxide you are able to tolerate, or your ventilatory response to carbon dioxide.

A strong ventilatory response to carbon dioxide means that your threshold will be reached sooner, resulting in a lower breath-hold time. Conversely, a good tolerance and reduced ventilatory response to carbon dioxide results in a higher breath-hold time.

When your BOLT score is lower, your breathing receptors are especially sensitive to carbon dioxide, and your breathing volume will be greater as the lungs work to remove any carbon dioxide in excess of programmed levels. However, when you have a normal tolerance to carbon dioxide and a higher BOLT score, you will be able to maintain calm breathing during rest and lighter breathing during physical exercise.

You may find that the first time you measure your BOLT score, you are surprised that your score is lower than expected—but remember that even elite athletes can have a low BOLT score! The good news is that your BOLT score can easily be increased with a series of simple breathing exercises incorporated into your existing way of life or exercise regimen. A common starting BOLT score for an individual who exercises regularly at a moderate intensity will be approximately 20 seconds. If your BOLT score is below 20 seconds, depending on genetic predisposition, you will probably find you experience a blocked nose, coughing, wheezing, disrupted sleep, snoring, fatigue, and excessive breathlessness during physical exercise. Each time that your BOLT score increases by 5 seconds, you will feel better, with more energy and reduced breathlessness during physical exercise. The aim of the Oxygen Advantage program is to increase your BOLT score to 40 seconds, and this can be realistically achieved.

Improving your BOLT score is an important key to attaining greater physical endurance. As we have already seen, having an improved tolerance to carbon dioxide means you are able to achieve a higher VO_2 max and improved performance. The Oxygen Advantage program is all about increasing your BOLT score and maximizing your potential!

How Your BOLT Score Relates to Breathlessness During Sports

The ideal BOLT score for a healthy individual is 40 seconds. In the book entitled *Exercise Physiology: Nutrition, Energy, and Human Performance* by William McArdle and colleagues, the authors observe: "If a person breath holds after a normal exhalation, it takes approximately 40 seconds before the urge to breathe increases enough to initiate inspiration."

What is accepted in theory is not always evident in practice. The truth is that the vast majority of individuals, including athletes, have a comfortable breath-hold time of about 20 seconds, often less. However, to achieve your full potential, a BOLT score of 40 seconds should be the goal.

Breath-hold measurements have also been used to study the onset and endurance of breathlessness (dyspnea) and asthma symptoms. The result that comes up again and again is that the lower the breath-hold time, the greater the likelihood of breathlessness, coughing, and wheezing during both rest and exercise.

Over the past thirteen years I have worked with thousands of children and adults with asthma. Although breath-hold time is not generally used by doctors to evaluate asthma severity, it is an excellent measurement to evaluate respiratory condition and symptoms such as coughing, wheezing, chest tightness, breathlessness, and exercise-induced asthma. If you experience breathlessness or asthma symptoms when you exercise, you may find that your athletic performance is limited and hampered by your condition. By implementing the Oxygen Advantage program and tracking your progress with your BOLT score, you will be able to quickly and easily improve your performance and eliminate symptoms of exercise-induced asthma. The overall goal of the Oxygen Advantage program is to increase your BOLT score to over 40 seconds, but every time you improve your BOLT score by 5 seconds you will find that symptoms such as coughing, wheezing, chest tightness, and breathlessness reduce drastically. More on eliminating exercise-induced asthma can be found in chapter 12.

BOLT Score and Breathing Volume

At this point, it can be useful to perform the following experiment:

- Sit down with a pen and paper.
- Bring your attention to your breathing and follow both the rate and depth of each breath.
- As you observe your breathing, draw the rate and depth across the sheet of paper.
- Do this for about half a minute or so, then check how your drawing relates to your BOLT score and the illustrations on the following page.

The following image is an example of the relationship between breathing volume and a BOLT score of 10 seconds.

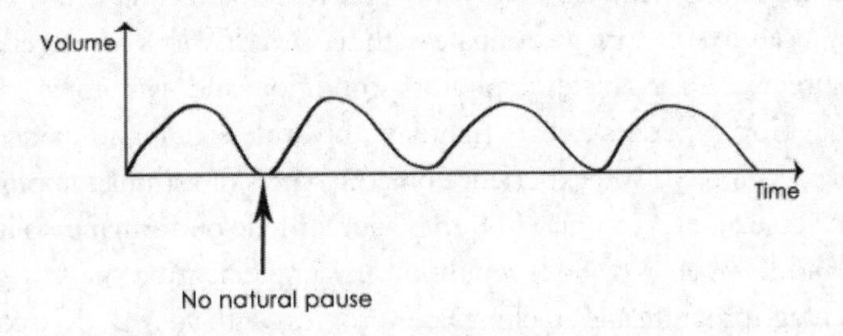

When a BOLT score is 10 seconds, breathing is noisy, loud, irregular, large, heavy, erratic, and effortful, with no natural pauses between breaths. If your BOLT score is 10 seconds or less, you will often experience a hunger for air, even when you are just sitting down. Habitual upper-chest breathing and mouth breathing is also expected. The number of breaths during rest per minute can be anything from 15 to 30 breaths.

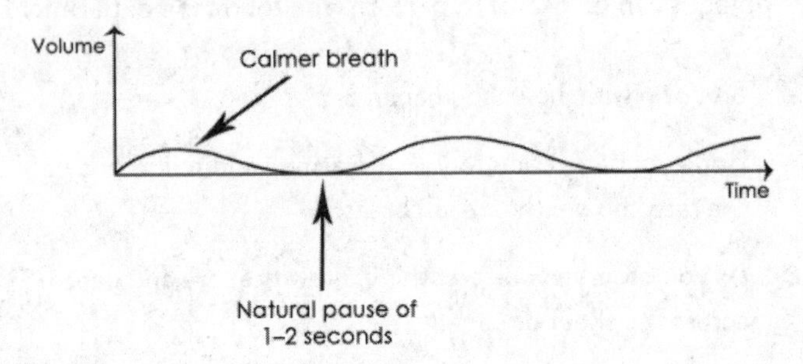

When a BOLT score is 20 seconds, breathing is heavy but regular. Both the rate and size of each breath is less than when a BOLT score is 10 seconds. A natural pause of between 1 and 2 seconds occurs at the end of each exhalation. The number of breaths per minute during rest will vary from between 15 and 20 moderately sized breaths.

BOLT 30 seconds

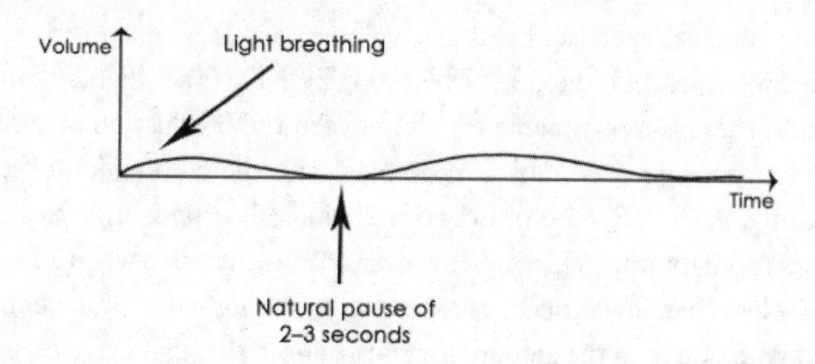

When a BOLT score is 30 seconds, breathing is calm, gentle, soft, effortless, and quiet. The rate and size of each breath continue to reduce as a BOLT score increases. The natural pause between each breath lengthens. The number of breaths during rest per minute will be about 10 to 15 minimal breaths.

BOLT 40 seconds

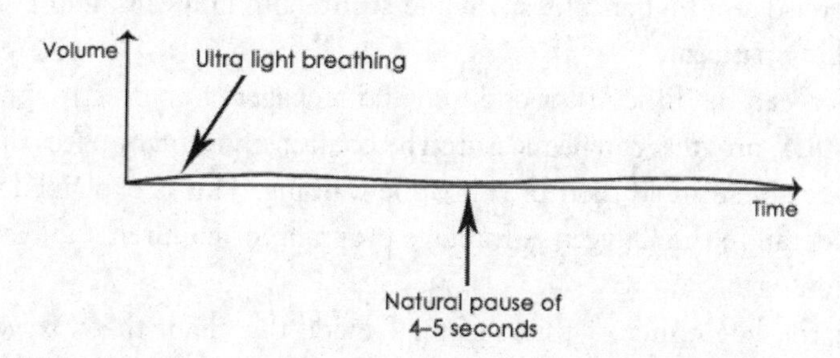

When a BOLT score is 40 seconds, breathing is effortless, calm, gentle, quiet, soft, and minimal. At a BOLT score of 40 seconds, it is difficult to see breathing movements. The natural pause between each breath is generally about 4 to 5 seconds. The number of breaths during rest per minute varies from 6 to 10 minimal breaths.

BOLT and Sports Performance

During physical exercise, breathing volume increases, but so too does production of carbon dioxide. When your BOLT score is higher than 30 seconds, there is a reasonable match between the production of carbon dioxide from increased muscle movement and the elimination of it by breathing. A BOLT score of between 20 and 30 seconds indicates that there is room for improvement. However, a problem arises when a BOLT score is less than 20 seconds, as excessive breathing will eliminate more carbon dioxide than the amount that is produced through exercise, leading to a net loss of CO_2, reduced oxygen delivery, and constriction of blood vessels and airways. The result is poor sports performance and a variety of negative implications for your overall health.

A general rule for the relationship between BOLT scores and breathing volume is as follows: The lower the BOLT score, the poorer the match between breathing volume and metabolic activity, hence the need to control breathing during rest and physical exercise. The closer the BOLT score is to 40 seconds, the better the match between breathing volume and metabolic requirements. When your breathing volume matches the amount of carbon dioxide produced, it will be much easier to exercise at a higher intensity while still retaining a calm and even breathing pattern.

As it can be difficult for children and teenagers to correctly apply the BOLT, progress can be measured by counting how many paces that they are able to hold their breath while walking. This is explained in more detail in the Oxygen Advantage program for children and teenagers on page 286.

At the beginning of this chapter I made the claim that Oxygen Advantage breathing exercises—even those that are practiced while sitting down—are guaranteed to increase your BOLT score. People with a BOLT score of less than 15 seconds are often put off from doing physical exercise due to excessive breathlessness. However, by starting with gentle reduced breathing exercises that can be performed while sitting down or walking slowly, even those with low BOLT scores can significantly improve their endurance and breathing efficiency in the space of just a few weeks. Once your BOLT score is greater than 20 seconds, you will be able to partake in physical exercise and continue to improve your score with more advanced breathing exercises.

Any athlete can learn the tools to tap into the body's own natural resources in order to train faster, exercise more intensively, and enjoy better improvements to his or her health. And if you are a coach, knowing the BOLT score of each of your players will help you to provide feedback on their ability to perform. As they say, knowledge is power; understanding your body's athletic capabilities will enable you to push harder and compete to the best of your abilities.

Three Steps to Increasing Your BOLT Score

The exercises in this book will guide you through each stage of your journey to improved breathing, fitness, and endurance. Below is a brief guide to each of the three steps to increasing your BOLT score:

1. Stop Losses of Carbon Dioxide

- Breathe through your nose, day and night.
- Stop sighing; instead, swallow or suppress the sigh. One sigh taken every few minutes is enough to maintain chronic overbreathing, so it is necessary to counteract the sigh by swallowing or holding the breath. If you notice your sighs only after they have taken place, then hold your breath for 10 to 15 seconds to help compensate for the loss of carbon dioxide.
- Avoid taking big breaths when yawning or talking. Individuals with a low BOLT score are often tired, and yawn frequently throughout the day. Try not to take in a large breath during a yawn. Likewise, individuals who talk for a living need to be aware that their breathing should not be heard during talking. If you find that you can hear

your breathing during talking, then it is better to slow down the speed of your talking, use shorter sentences, and take a gentle breath through your nose between each sentence.

- Observe your breathing throughout the day. Good breathing during rest should not be seen or heard.

2. Improve Tolerance of Carbon Dioxide

This is where you practice exercises designed to reduce your breathing volume toward normal. They will bring a feeling of relaxation to your body and encourage your breathing to slow down and become calmer. The objective is to create a tolerable need or hunger for air. A sustained need for air over the course of 10 to 12 minutes resets the receptors in the brain to tolerate a higher concentration of carbon dioxide.

Steps 1 and 2 are necessary to increase BOLT score from 10 to 20 seconds.

3. Simulate High-Altitude Training

During physical exertion, as discussed, breathing volume increases along with metabolic activity, which generates carbon dioxide. Breathing less than you feel you need to during physical training is an excellent method of conditioning the body to tolerate a higher concentration of carbon dioxide, while at the same time subjecting the body to a reduced concentration of oxygen.

The benefit of implementing Oxygen Advantage breathing techniques during physical exercise is that a stronger air shortage can be created than at rest. A strong air shortage is necessary to increase your BOLT score from 20 to 40 seconds.

Please note the following important points about increasing your BOLT score:

- You will feel better *each time* your BOLT score increases by 5 seconds.
- The general progression is for a BOLT score to increase by 3 to 4 seconds during the first 2 to 3 weeks. When your BOLT score reaches 20 seconds, it is normal for the progression of your BOLT increase to slow down. It is not uncommon for a BOLT score to remain "stuck" at 20 seconds for 8 to 10 weeks. In order to increase a BOLT score from 20 to 40 seconds, it is necessary to perform physical exercise while incorporating the techniques in this book. Be determined, and don't lose heart if your BOLT score is stubborn or lessens temporarily! Meanwhile, you'll enjoy the benefits you've already gained from getting to 20 seconds in the first place.
- The causes for slow BOLT score progress include lifestyle factors, such as stress and excessive talking, and sickness. The severity and duration of a health condition will dictate the rate of your BOLT progress, but whatever your current state of health, there are always exercises that can be performed to keep you moving forward. It is worth persevering, as there are many health benefits related to even a slight improvement to your BOLT score.
- The most accurate BOLT score is taken first thing after waking. This BOLT measurement is more accurate because you cannot influence your breathing during sleep, and therefore an early morning score will be based on your breathing volume as naturally set by your respiratory center.
- Your goal is to maintain a morning BOLT score of 40 seconds for a period of 6 months. As modern living can negatively affect BOLT scores, it is necessary to pay attention to breathing throughout the day, ensuring it

is light and through the nose, and to incorporate the Oxygen Advantage program into your physical exercise and way of life. This will help you maintain a high BOLT score.

Below are suggested exercises as based on your BOLT score (a more detailed program is available on page 267):

BOLT score of 10 seconds or under:

- Breathing Recovery Exercise to unblock the nose
- Nose-breathe at all times
- Avoiding sighing and taking big breaths
- Breathe Light to Breathe Right during rest
- Breathing Recovery Exercise

BOLT score of 10 to 20 seconds:

- Nose unblocking exercise
- Nose-breathe at all times
- Avoiding sighing and taking big breaths
- Breathe Light to Breathe Right during rest and physical exercise

BOLT score of 20 to 30 seconds:

- Nose unblocking exercise
- Nose-breathe at all times
- Breathe Light to Breathe Right during rest and physical exercise
- Simulate High-Altitude Training during a fast walk or jog

BOLT score of 30 seconds or greater:

- Nose-breathe at all times
- Breathe Light to Breathe Right during rest and physical exercise
- Simulate High-Altitude Training during jogging or running
- Advanced Simulation of High-Altitude Training

To reiterate: If you have any health problems or a BOLT score shorter than 10 seconds, please do not attempt breath holds involving a strong need for air, as the resultant loss of control of your breathing may aggravate your condition. Please do not attempt the nose unblocking exercise or any of the exercises to simulate high-altitude training unless you have a BOLT score longer than 10 seconds. It is also advised to have a BOLT score of at least 20 seconds before attempting breath holding during jogging or running.

Body Detoxification

When you undergo the Oxygen Advantage program, you may experience a body detoxification. The extent of the detoxification will depend on your BOLT score and state of health. In general, the higher your BOLT score and the healthier you are, the less likely you will experience a detoxification. On the other hand, if you have a low BOLT result and have been feeling unwell for years, you are much more likely to experience a detoxification. Remember that changing your breathing volume toward normal improves blood flow and oxygenation of all tissues and organs. With better-functioning organs and systems, waste products are expelled more readily.

It is great news if you do experience a detoxification, as it shows your health is improving, and you should clearly feel the benefits. Generally, cleansing reactions are mild and last from several hours to one or two days.

Typical detoxification symptoms may include:

- Increased demand for water
- Loss of appetite
- Bad taste in the mouth
- Increased moodiness
- Short-term headache
- Increased secretions of mucus from the lungs by people with asthma
- Head cold with runny nose, especially during physical exercise
- Diarrhea

An integral part of detoxification is a reduced appetite for food, so it's important to only eat when hungry. To help reduce the intensity and duration of any cleansing reactions, drink warm water regularly throughout the day and continue with reduced breathing exercises.

In the next section we will begin the first step to improving your BOLT score: nasal breathing. We will look at the functions of the nose, how to decongest it, and learn what nasal breathing offers for health, sports . . . and in the bedroom.

CHAPTER 3

Noses Are for Breathing, Mouths Are for Eating

In order to address breathing volume and increase BOLT score, the first step is to go back to basics and learn to breathe through the nose both day and night. As any child is aware, our nose is made for breathing, the mouth for eating. You were born breathing through your nose, and it has been our primary conduit for breathing for hundreds of thousands of years.

It was only when our ancient ancestors were in dangerous situations that they reverted to mouth breathing to take in greater volumes of air in preparation for intense physical activity.

It is for this reason that mouth breathing is synonymous with emergency, activating the same fight-or-flight response that our ancestors experienced but these days usually without the accompanying physical exercise to allow our operating systems to revert to normal. From the perspective of breathing physiology, mouth breathing activates use of the upper chest, while nasal breathing results in abdominal breathing. You can verify the difference by sitting in front of a mirror and placing one hand on your chest and one hand above your navel. Once settled, take a moderate-size breath in through your mouth and note the movements of your hands. Next, compare your breathing movements to a similar size breath drawn in through your nose.

Upper-chest breathing is more likely to be associated with a stress response, while nasal breathing helps ensure regular, calm, steady breathing using the diaphragm. The common misconception of taking a "deep" breath is to puff out the chest and raise the shoulders, but this is neither deep nor beneficial to oxygenating the body. To help deal with stress, the instruction to take a deep breath is actually correct, but a truly deep breath is abdominal, gentle and quiet; the exact opposite of the big breaths usually taken in an attempt to calm down.

Mouth breathing activates the upper chest, involves larger breaths, and may cause reduced oxygen uptake in the arterial blood. It is no wonder that habitual mouth breathers often suffer from poor energy, a lack of concentration, and moodiness. We all know the stereotype of the mouth breather, portrayed by moviemakers from Hollywood to Bollywood as an idiot. But in case you think I am being unduly critical, I was a mouth breather for more than twenty years, so I know all too well the effects. Furthermore, every time I look in the mirror I see the results from my years of mouth breathing. Dentists and orthodontists have also documented these profound facial changes as a result of habitual mouth breathing: narrow jaws, crooked teeth, sunken cheekbones, and smaller nasal cavities. While orthodontic treatment and the wearing of braces are epidemic among modern-day teenagers, it was normal for our ancestors to have wide faces with perfectly shaped teeth.

In the 1930s, a dentist by the name of Dr. Weston Price investigated the cause of facial changes and crooked teeth in various countries and civilizations. One of his observations while visiting Gaelic people living on the Hebridean islands off the coast of Scotland was that children became mouth breathers after parents switched from their natural diet of seafood and oatmeal to the modernized diet of "angel food cake, white bread and many white flour commodities, marmalade, canned vegetables, sweetened fruit juices, jams, and confections."

Dr. Price's discovery illustrates the link between modern diet and chronic hyperventilation. Processed foods are mucus and acid forming. Throughout evolution, our diet consisted of 95 percent alkaline-forming and 5 percent acid-forming foods. Nowadays the reverse is true: Our diet is 95 percent acid- and 5 percent alkaline-forming foods. Acid-forming foods—such as processed products, dairy, meat, bread, sugar, coffee, and tea—stimulate breathing. A natural response to experiencing a greater demand to breathe is to open the mouth to take in more air. Over time, the brain adjusts to this larger intake of air, and overbreathing becomes a habit.

On the other hand, alkaline-forming foods such as fruit and vegetables, along with plain water, are easy for the body to process; they are "breathing-friendly" foods. But while these types of foods are highly beneficial, I'm not saying you need to become a vegetarian. Protein is an essential part of a healthy diet, and meat provides a natural, nutrient-rich source. The most important change is to get rid of processed foods in your diet. They may take up the most space in our supermarkets, but they are effectively suitable food for no one.

The nose is one of the most important organs in the human body. In his nineteenth-century travels in North America, the artist George Catlin noticed that the Native American mothers paid a lot of attention to their infants' breathing. If at any time the baby opened its mouth to breathe, the mother would gently press the baby's lips together to ensure continued nasal breathing. Catlin also noted that the rate of sickness and illness among the native Indian people was very low in comparison with European settlers. In his aptly titled 1882 book *Shut Your Mouth and Save Your Life,* Catlin wrote, "When I have seen a poor Indian woman in the wilderness, lowering her infant from the breast, and pressing its lips together as it falls asleep . . . I have said to myself, 'Glorious education! Such a mother deserves to be the nurse of Emperors.' " In comparison, Catlin described how the babies of the European settlers slept with their mouths open, gasping for breath in stuffy, hot, and unventilated rooms.

Nasal breathing is often an integral part of an animal's survival or hunting techniques. The cheetah, which is considered the fastest land animal on earth, is capable of accelerating from 0 to 60 miles per hour in just 3 seconds. Most high-performance cars cannot accelerate so quickly, with the notable exception of the Bugatti Veyron, which will set you back a million dollars to experience the natural acceleration of a cheetah. With such incredible efficiency and speed, it doesn't take long for the cheetah to catch up with its prey, but maintaining nasal breathing is especially advantageous during the chase, ensuring that its victim is the first to run out of air.

The dog is probably the best-known example of an animal that periodically breathes through its mouth—dogs can commonly be seen panting on a hot day or after a long walk to help cool themselves down. But at all other times a dog will breathe through its nose, only using its mouth for eating, drinking, and barking. Nature has ensured that the vast majority of land mammals breathe through their noses by positioning the windpipe so that the back of the nose leads directly to the lungs. In other words, it is not easy for most animals to breathe through their mouths.

The same is true for humans at birth, but after a few months the windpipe drops down to just below the back of the tongue in order to allow the baby to breathe through both its mouth and nose. Charles Darwin was puzzled by this adaptation in humans: how, unlike most animals, the openings for carrying food to the stomach and air to the lungs are placed side by side. This parallel position seems fairly impractical, as it increases the risk of food going down the wrong way, requiring the development of a complicated swallowing mechanism. The cause for this is likely to do with our ability to speak and to enable us to swim, since both actions require voluntary control over breathing. Had Darwin investigated the negative impact of mouth breathing in human beings, however, I have no doubt that he would have considered the ability to mouth breathe to be a far worse flaw in the evolution of our species than the risk of choking while eating.

The rest of the animal kingdom relies on nasal breathing for survival, and mouth breathing usually only occurs as an adaptation within a species. Birds, for example, are predominantly nose breathers, aside from diving birds such as penguins, pelicans, or gannets. Generally, when an animal breathes through its mouth it is a sign of sickness, injury, or distress. Guinea pigs and rabbits will continue to breathe through their noses even under heavy exertion and will only breathe through their mouths if they have developed a breathing abnormality.

The same goes for all farm animals, including the cow, sheep, donkey, goat, and horse. Mouth breathing in these animals would be a clear signal to a farmer or pet owner that there is something wrong. Experience tells the farmer that when a cow or sheep stands motionless with its neck extended and mouth open, it is very sick—time to call the vet.

When it comes to the importance of breathing through the nose, there is no distinction between prey and predator. Nasal breathing is especially advantageous for horses and deer, since it allows them to graze and breathe at the same time, while their sense of smell alerts them of approaching predators. I, on the other hand, as a habitual mouth breather, was admonished as a child for chewing with my mouth open—much to the dismay of my well-mannered table companions. Unlike the horse, I could not eat and breathe through my nose at the same time. And, unlike the horse, I found myself completely out of breath after attempting even light exercise—meanwhile, spend a day at the horse track and you will witness these majestic animals running at speeds of up to thirty miles per hour while maintaining nose breathing.

To get an idea of the size of the nasal cavity, run your tongue from the front of the roof of your mouth right back as far as it will go. You may be surprised to learn that the roof of the mouth is in fact the floor of the nose! The nose you see on your face comprises approximately 30 percent of its volume. It is the tip of the iceberg, so to speak, with the remaining 70 percent of the nasal cavity set deep within the skull. Nature is intelligent and does not waste space; evolution has determined the importance of the nose by the amount of space it occupies within the skull.

As air enters through the nose, it is swirled through scrolled, spongy bones called *turbinates,* which condition and guide inhaled air into a steady, regular pattern. The internal nose, with its cul de sacs, valves, and turbinates, regulates the direction and velocity of the air to maximize exposure to a network of small arteries and veins and to the mucous blanket in order to warm, humidify, and sterilize the air before it is drawn to the lungs. The late Dr. Maurice Cottle, who founded the American Rhinologic Society in 1954, stated that the nose performs at least thirty functions, all of which are important supplements to the roles played by the lungs, heart, and other organs. The large amount of space in the skull devoted to the nasal cavity provides an indication of the importance of the functions of the nose.

To attain a higher BOLT score and improved sports performance, it is imperative that nasal breathing is practiced at all times during rest. If your BOLT score is less than 20 seconds, the only way to avoid overbreathing during exercise is to breathe through the nose at all times, even while training. An exception to nasal breathing can be made for a short period of time during intensive physical exercise, but this kind of training should only be attempted when your BOLT score is greater than 20 seconds.

The Nose—A Most Important Organ

In the yoga book *The Science of Breath,* written over a century ago, Yogi Ramacharaka said this about nostril versus mouth breathing: "One of the first lessons in the Yogi Science of Breath is to learn how to breathe through the nostrils, and to overcome the common practice of mouth breathing." It seems that little has changed over the past hundred years—if anything, the prevalence of mouth breathing has increased. Extolling the benefits of nasal breathing, Yogi Ramacharaka surmises that "many of the diseases to which civilized man is subject are undoubtedly caused by this common habit of mouth breathing." Below is a brief list of the functions of nasal breathing:

- Nose breathing imposes approximately 50 percent more resistance to the airstream in normal individuals than does mouth breathing, resulting in 10 to 20 percent more O_2 uptake.
- Nasal breathing warms and humidifies incoming air. (Air entering the nose at 42.8°F/6°C will be warmed to 86°F/30°C by the time it touches the back of the throat,

and a cozy 98.6°F/37°C—body temperature—upon reaching its final destination, the lungs.)

- Nasal breathing removes a significant amount of germs and bacteria from the air you breathe in.
- Nasal breathing during physical exercise allows for a work intensity great enough to produce an aerobic training effect as based on heart rate and percentage of VO_2 max.
- As discussed in the next section, the nose is a reservoir for nitric oxide, an essential gas for the maintenance of good health.

Now compare the benefits above with the effects of mouth breathing:

- Mouth-breathing children are at greater risk of developing forward head posture, and reduced respiratory strength.
- Breathing through the mouth contributes to general dehydration (mouth breathing during sleep results in waking up with a dry mouth).
- A dry mouth also increases acidification of the mouth and results in more dental cavities and gum disease.
- Mouth breathing causes bad breath due to altered bacterial flora.
- Breathing through the mouth has been proven to significantly increase the number of occurrences of snoring and obstructive sleep apnea.

The Nose: A Great Source for Nitric Oxide

Until the 1980s the gas nitric oxide (NO) was considered a toxic substance, causing smog and producing harmful effects in the environment. When the first article appeared discussing the importance of nitric oxide, the scientific community found it difficult to conceive that a gas so toxic outside of the body could play such a highly important role within it. And although nitric oxide is a relative newcomer to the field of medicine, there are now over one hundred thousand research papers devoted to this gas, providing an insight into how it has captured the attention of doctors and scientists alike.

In 1992, nitric oxide was proclaimed Molecule of the Year by the journal *Science* and was described as a startlingly simple molecule that unites neuroscience, physiology, and immunology and revises scientists' understanding of how cells communicate and defend themselves.

In 1998, Robert F. Furchgott, Louis J. Ignarro, and Ferid Murad were awarded the Nobel Prize for their discovery that the gas nitric oxide is an important signaling molecule in the cardiovascular system. When I first began to read about the benefits of nitric oxide, I was astounded as to how one simple gas could influence all major systems and organs, help keep us free from disease including cancer, promote a longer life, and even perform better in bed.

Strangely, despite these life-changing attributes, it seems that few people outside of the field of medicine are aware of this gas and its tremendous benefits to health. Of the hundreds of people I have spoken to with high blood pressure, poor cardiovascular health, asthma, and other ailments, not one of them was aware of the importance of nitric oxide.

When it comes to nasal breathing and breath-hold exercises, nitric oxide plays an important role. Nitric oxide is produced inside the nasal cavity and the lining of the thousands of miles of blood vessels throughout the body.

Scientific findings have shown that this extraordinary molecule is released in the nasal airways and transferred to the lower airways and lungs through nasal breathing. In the respected medical journal *Thorax,* researchers Jon Lundberg and Eddie Weitzberg from the world-famous Karolinska Institute in Sweden state, "Nitric oxide (NO) is released in the nasal airways in humans. During inspiration through the nose, this NO will follow the airstream to the lower airways and the lungs."

Recognizing the importance of the role of nitric oxide in oxygenation of the body, Dr. Mehmet Oz recommends inhaling from the diaphragm as it "brings nitric oxide from the back of your nose and your sinuses into your lungs. This short-lived gas dilates the air passages in your lungs and does the same to the blood vessels."

Nasal breathing is imperative for harnessing the benefits of nitric oxide, working hand in hand with abdominal breathing and helping to maximize body oxygenation. Think of the nose as a reservoir: Each time we breathe gently and slowly through the nose, we carry this mighty molecule into the lungs and blood, where it can do its work throughout the body. Mouth breathing bypasses this special gas, missing out on the important advantages that nitric oxide provides for general well-being.

Nitric oxide plays an important role in vasoregulation (the opening and closing of blood vessels), homeostasis (the way in which the body maintains a state of stable physiological balance in order to stay alive), neurotransmission (the messaging system within the brain), immune defense, and respiration. It helps to prevent high blood pressure, lower cholesterol, keep the arteries young and flexible, and prevent the clogging of arteries with plaque and clots. All these benefits reduce your risk of heart attack and stroke—two of the top three killers in America.

As we age, blood vessels lose flexibility and reduce blood circulation throughout the body. It is no coincidence, therefore, that as men grow older, conditions related to reduced blood flow—including erectile dysfunction—become more prevalent. The potency of nitric oxide in opening blood vessels becomes clear when you realize that this simple gas plays a significant role in erection of the penis. This discovery, in fact, led to the production in 1998 of Viagra, a very popular drug that received thousands of hours of free media time and generated sales of billions of dollars for the manufacturer, Pfizer.

There are many causes of habitual mouth breathing, including swelling of the tissue in the nose to form nasal polyps. In a study of a group of thirty-three men with nasal polyps, the authors found that erectile dysfunction was significantly higher in this group. Furthermore, when the men underwent surgery to remove the polyps and allow restoration of nose breathing, erectile dysfunction was significantly ameliorated.

And women can benefit from nitric oxide in this way too, as the gas plays a similar role in the female genitalia, helping to increase libido. Could it be that nose breathers have more desire and better sex lives than mouth breathers?

In addition to improving your sex life, this unique gas also acts as a defense mechanism against microorganisms through its antiviral and antimicrobial activity, potentially reducing the risk of illness and improving overall health.

Most important for athletes wishing to optimize their sports performance, nitric oxide plays a central role in dilating the smooth muscle layer embedded in the airways. Open airways allow for a better transfer of oxygen to and from the lungs during exercise, while tight airways create an uncomfortable and inefficient experience that ultimately affects performance.

The production of nitric oxide in the nasal sinuses can be increased by simply humming. In an article published in the *American Journal of Respiratory and Critical Care Medicine,* Doctors Weitzberg and Lundberg described how humming increased nitric oxide up to fifteenfold in comparison with quiet exhalation. They concluded that humming causes a dramatic increase in sinus ventilation and nasal nitric oxide release.

With this knowledge, it comes as no surprise that humming is also practiced during certain meditation techniques. The breathwork technique called Brahmari involves slow, deep breaths through the nose, humming on each exhalation to generate a sound similar to a bee buzzing, and while the exact science may have been a mystery to the creators of this meditation method, the associated feeling of calmness of the mind is a clear indication of its benefit.

Nose Unblocking Exercise

Breathing through the mouth causes blood vessels in the nose to become inflamed and enlarged. This, along with an increased secretion of mucus, creates the uncomfortable feeling of nasal stuffiness. When the nose becomes blocked it is much more difficult to breathe through it, thus perpetuating the habit of breathing through the mouth. Continued mouth breathing results in a more permanent state of nasal congestion, thus completing the vicious circle.

Nasal obstruction is one of the main symptoms of rhinitis and affects many people throughout the Western world on a daily basis. The most common treatments include the avoidance of triggers (such as pollen) and the use of decongestants, nasal steroid sprays, antihistimines, or allergy shots, but while these offer symptomatic benefits, they are effective only as long as treatment continues.

A number of years ago, ear, nose, and throat specialist Professor John Fenton from the University of Limerick took an interest in my work after his patients reported a significant reduction in their nasal symptoms following attendance of my course. At his behest, a study was commissioned to further investigate the effects of reduced breathing. The results were an amazing 70 percent reduction of symptoms such as nasal stuffiness, poor sense of smell, snoring, trouble breathing through the nose, trouble sleeping, and having to breathe through the mouth.

On the following page is one of the exercises that I taught to participants in the study. (Please do not practice this exercise if your BOLT score is less than 10 seconds, or if you have high blood pressure or other cardiovascular issues, diabetes, or pregnancy, or have any other serious health concerns.) Like all breathing exercises, the Nose Unblocking Exercise should not be practiced right after eating.

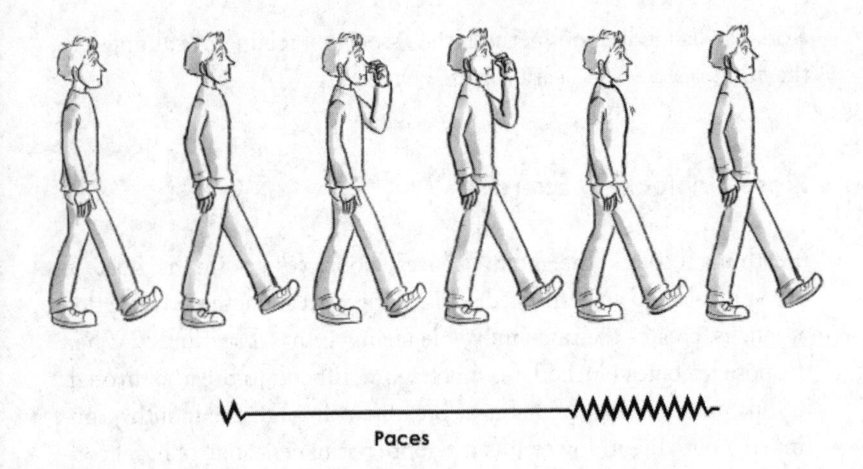

Nose Unblocking Exercise

- Take a small, silent breath in through your nose and a small, silent breath out through your nose.
- Pinch your nose with your fingers to hold your breath.
- Walk as many paces as possible with your breath held. Try to build up a medium to strong air shortage, without overdoing it.
- When you resume breathing, do so only through your nose. Try to calm your breathing immediately.
- After resuming your breathing, your first breath will probably be bigger than normal. Make sure that you calm your breathing as soon as possible by suppressing your second and third breaths.
- You should be able to recover normal breathing within 2 or 3 breaths. If your breathing is erratic or heavier than usual, you have held your breath for too long.

- Wait 1 or 2 minutes before repeating the breath hold.
- In order to prepare yourself for the longer breath holds, go easy for the first few repetitions, increasing your paces each time.
- Repeat for a total of 6 breath holds, creating a fairly strong need for air.

Generally, this exercise will unblock the nose, even if you have a head cold. However, as soon as the effects of the breath hold wear off, the nose will likely feel blocked again. By gradually increasing the number of steps you can take with your breath held, you will find the results continue to improve. When you are able to walk a total of 80 paces with the breath held, your nose will remain decongested. Eighty paces is actually a very achievable goal, and you can expect to progress by an additional ten paces per week.

Each week I teach this exercise to groups of five- to ten-year-old children, many of whom have pretty serious breathing difficulties. Within 2 or 3 weeks, most children are able to walk 60 paces with their breath held, with some children quickly achieving up to 80 paces. Try it yourself, and see how you do.

If you regularly suffer from nasal congestion, you should soon find it much easier to breathe through your nose by practicing this exercise. No longer will you require over-the-counter nasal decongestants, antihistamines, or nasal steroids!

By holding your breath, you sharply increase the concentration of nitric oxide in your nasal cavity, resulting in dilation of the nasal passages and smooth, easy nasal breathing once more.

As you move on to the breathing exercises in the next chapter, your ability to hold your breath will improve, resulting in even greater nasal freedom.

Nasal Breathing at Night

The ideal amount of sleep required each night varies from person to person. The late British prime minister Margaret Thatcher was said to get by on just four hours sleep, but most of us need seven or eight hours of good-quality sleep to set us right for the day. If we struggle to get to sleep, or if sleep is interrupted by snoring or sleep apnea, it can be particularly hard to get up the following morning, and a lack of sleep can severely affect concentration, mood, and even the most basic of activities. Even if we seem to sleep soundly the whole night through, the quality of sleep can be reduced by mouth breathing and heavy breathing, meaning we still wake up with a dry mouth and a feeling of lethargy.

Fifty-year-old Annette described to me how she rarely slept eight hours continuously. Since her children were very young, her usual sleep pattern had involved lying awake for several hours in an effort to try to fall asleep, followed by a few hours of light sleep, waking again around 3 A.M. The next two hours were usually spent trying to fall asleep again, resulting in frustration and exhaustion when she finally had to get up for her day's work.

Just like Annette, for many years I too woke up tired and lethargic, suffering from poor concentration throughout the day. The key to improving the quality of my sleep was incredibly simple: All I had to do was to learn to keep my mouth closed during sleep. Because we are unaware of how we breathe at night, the only sure way to ensure nasal breathing is to wear light paper tape across the lips to prevent the mouth from falling open. And this is exactly what I instructed Annette to do. If you feel uncomfortable about using the paper tape at night, a stop snoring strap is a good alternative and helps to keep the lower jaw from dropping during sleep. Stop snoring straps are commonly used by people with obstructive sleep apnea and can be purchased from OxygenAdvantage.com.

Taping the mouth at night ensures the benefits of good breathing during sleep, allowing you to fall asleep more quickly, stay asleep longer, and wake feeling energized. The tape that I have found most suitable, as it is simple to use, hypoallergenic, and light, is 3M Micropore tape, which can be bought from most drugstores. To help make the tape easier to remove in the morning, gently press the tape against the back of your hand a couple of times to remove some of the glue before applying the tape to your face. All you need to do is tear off about 4 inches/10 cm of tape, fold a tab over at both ends to make removal easier in the morning, dry your lips, close your mouth, and gently place the tape horizontally over the lips.

At first, Annette was slightly nervous about using the tape, though she was keen to try anything that might help her sleep and increase her energy levels. Initially, she found taping her mouth to be uncomfortable, and noticed that her breathing increased due to anxiety when she used it. However, over the next few days she practiced wearing the tape for short periods of 20 minutes while she went about her normal activities at home. This helped to acclimatize her to breathing through her nose and to overcome any fears about wearing the tape at night.

Once she was used to the feeling of taping her mouth closed, Annette was determined to use the technique to improve her sleep quality. She went to bed at her usual time and was surprised to find the tape to be a comfort. As soon as she placed the tape on her mouth, it was almost like a signal to fall asleep—which she did. That night, Annette slept more deeply than usual, and even though she woke up without the tape on the first two days, she felt more rested. On her third day, she went to bed wearing the tape at 10 P.M. and slept like a baby until 9:53 A.M. Annette excitedly told me that it was the first time in years that she'd had a proper night's sleep, and she was amazed to feel so alert and energized when she woke up.

Over the years, I have introduced this taping method to thousands of people with incredible results. Unless you breathe calmly through your nose at night, you have no idea what it feels like to have a great night's sleep. Taping the mouth at night is a simple but very effective technique, and while it may sound a little strange, it is well worth getting used to.

Continue to wear the tape until you have managed to change to breathing through your nose at night. How long this takes will vary from person to person, but in general wearing the tape for a period of around three months is sufficient to restore nasal breathing during sleep. Breathing through your nose will result in a naturally moist mouth when you wake up. If your mouth is dry upon waking, you know that your mouth was open during sleep. When a child has one eye with weaker vision, the treatment often recommended is to temporarily cover the good eye with a patch to train the brain to strengthen the weaker eye and restore normal vision. In the same way, wearing tape across the lips during sleep or when alone in your house during the day gradually trains the body to adapt to nasal breathing both day and night. Spending a guaranteed eight hours breathing through your nose while you sleep is an opportune way to reeducate your respiratory center to adjust to a more normal breathing volume.

CHAPTER 4

Breathe Light to Breathe Right

For thousands of years, masters of the ancient arts of yoga, tai chi, and qigong have espoused the importance of quiet, gentle, and light breathing. I recently had the pleasure of meeting tai chi Master Jennifer Lee in London. Master Lee has reached the rank of Seventh Dan and was awarded gold in ten categories during the international 2009 Wushu Championships held in Hong Kong and Hainan, China. As the two of us talked, Master Lee described the similarities between her work and mine. She explained that during tai chi tournaments, judges pay particular attention to whether they can notice the breathing of competitors, with points being deducted when breathing is evident.

Without knowing why, other than the fact that it has been passed down from generation to generation, Master Lee practices a breathing exercise that is very similar to the reduced breathing exercise we will explore further on. It is no coincidence that Master Lee's breathing was textbook perfect. It was abdominal, effortless, and almost invisible to the eye. I have watched many people breathe—thousands, in fact—and without doubt, Master Lee displayed the most perfect breathing I have ever seen.

Well-known qigong and tai chi Master Chris Pei explains how breathing is at the very core of the Chinese concept of *chi* (*qi*): "Generally speaking, there are three levels of breathing. The first one is to breathe *softly,* so that a person standing next to you does not hear you breathing. The second level is to breathe softly so that *you* do not hear yourself breathing. And the third level is to breathe softly so that you do not *feel* yourself breathing."

This philosophy of effortless breathing is echoed by authentic teachers of Indian yoga and traditional Chinese medicine. I use the word *authentic* in order to differentiate practitioners who have a deep knowledge of breathing and how it affects physiology from those who don't. Unlike many modern Western teachers of yoga, who instruct students to breathe hard in order to remove toxins from the body, authentic teachers know that when it comes to breathing, less is more. The traditional Chinese philosophy of Taoism succinctly describes ideal breathing as "so smooth that the fine hairs within the nostrils remain motionless." True health and inner peace occurs when breathing is quiet, effortless, soft, through the nose, abdominal, rhythmic, and gently paused on the exhale. This is how human beings naturally breathed until modern life changed everything.

The comedian Lavell Crawford—who is, admittedly, quite a large man—got a laugh from his audience while telling the story of a young boy who stopped him in the street to ask: "You are so big—how many stomachs do you have?" This inquiry was quickly followed by, "Why are you breathing so hard, you got asthma or something?" While the kid's manners left something to be desired—a fact that Crawford observed in his act—he got it right when he recognized the problematic nature of noticeable breathing.

The commonly used practice of taking big breaths is based on the misconception that taking in more air will increase the oxygen levels of the blood. However, since arterial blood is already almost fully saturated with oxygen (between 95 percent and 99 percent) during normal, healthy breathing, "big" breathing is rendered totally unnecessary.

Authentic teachers are not adding anything new. Instead, they are helping to counteract the negative effects imposed on breathing by processed foods, stress, excessive talking, stuffy air, and the false belief of the benefits of taking big breaths. Authentic professional yoga practitioners will have developed a high tolerance to carbon dioxide through their practice—sometimes to the point of being able to sustain one calm breath per minute for a whole hour! This impressive breathing efficiency implies quiet, gentle breathing and a high BOLT score. This is exactly the goal of the Oxygen Advantage program: to bring your breathing back to basics and to incorporate the wisdom and time-tested principles of ancient man that are enshrined by authentic teachers.

What Is a Deep Breath? Unveiling the Myth

Sometimes the same word can conjure up different meanings for different people. Take the word *deep* as an example: "deep" can be defined as a measurement that "extends far from the top," but this explanation is not always clear-cut. The bottom of the deep end of a swimming pool is obviously farther from the surface of the water than the bottom of the shallow end, but when the word *deep* is used in the context of taking a deep breath, it can be open to many different interpretations. Such an instruction is commonly given by stress counselors, yoga practitioners, and sports coaches, resulting in the student taking a big breath of fresh air into the lungs, but often this is done through an open mouth while activating the upper chest. Such breathing is both big and shallow, but not deep. This type of breathing is entirely the wrong thing to do if your objective is to deliver more oxygen throughout the body.

If we use the definition of "extending far from the top" in the context of deep breathing, the "top" will refer to the top of the lungs or the upper chest. A deep breath, therefore, means to breathe down into the full depth of the lungs. It also means using the main breathing muscle, the diaphragm, which separates the chest from the abdomen. During rest, healthy animals and babies naturally take deep, silent breaths. With each inhalation and exhalation the abdomen gently expands and contracts. There is no effort involved; the breath is silent, regular, and, more important, through the nose. If you want to learn what constitutes good breathing, observe the breathing of a baby or a healthy pet, whose breathing has not been altered by the effects of modern lifestyles.

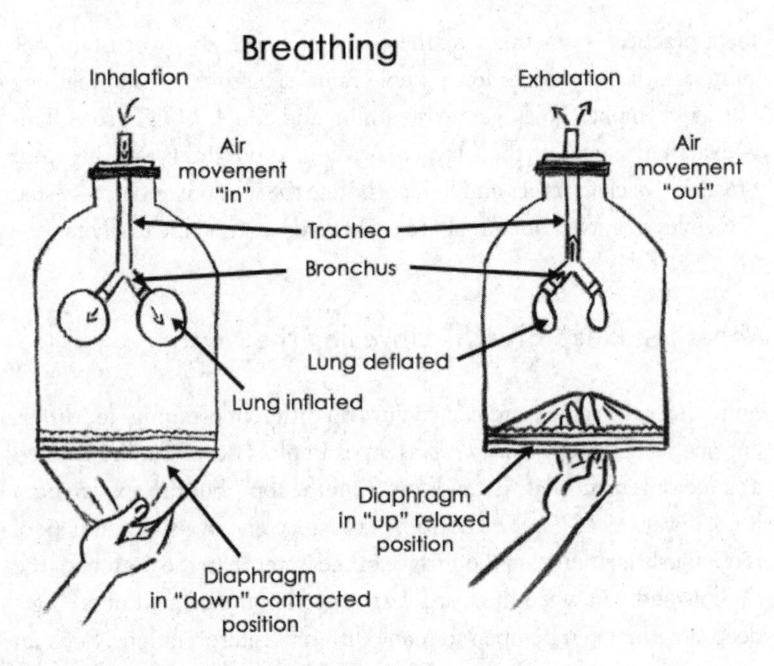

Bell jar experiment showing "how we breathe"

The bell jar diagram above illustrates the relationship between the movement of the diaphragm and movement of the abdomen. As you breathe, your abdomen moves in the opposite direction. The reason why the abdomen moves outward with the in-breath is because the diaphragm pushes downward, exerting gentle force on the abdomen. As you exhale, the diaphragm moves upward, taking pressure off the abdomen and moving the abdomen inward.

To bring air down into the depths of the lungs, it is not actually necessary to take a big breath, as even the quietest of breaths will activate the diaphragm. When you are practicing abdominal nasal breathing, you should not be able to see or hear your breath during rest. In contrast, overbreathing through the mouth, in an attempt to take a "deep" breath, will be clearly audible and cause the chest to rise and fall, but will still not manage to draw the breath deeply into the lungs.

The Diaphragm

The diaphragm is a dome-shaped sheet of muscle that separates the thorax (which houses the heart and lungs) from the abdomen (which houses the intestines, stomach, liver, and kidneys). The diaphragm serves as our main breathing muscle and, if used correctly, it provides deep and efficient breathing. Poor breathing habits do not take full advantage of the diaphragm and instead encourage inefficient overbreathing from the upper chest. To determine where your diaphragm is located, place your hands just at the base of your rib cage and follow your ribs from the front of your body to the sides. A good rule of thumb is that the diaphragm is located about four buttons down on your shirt.

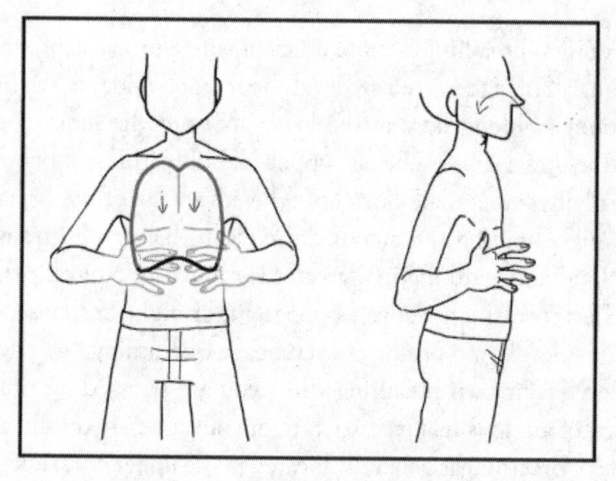

Breathing in (inhaling)—abdomen gently moves outward

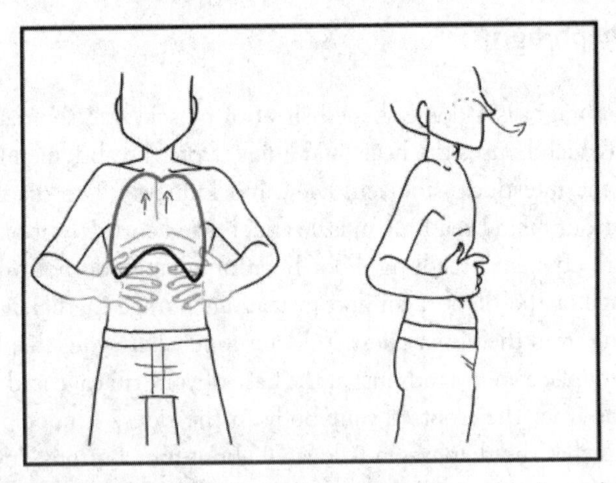

Breathing out (exhaling)—abdomen gently moves inward

Abdominal breathing is more efficient simply because of the shape of the lungs. Since they are narrow at the top and wider at the bottom, the amount of blood flow in the lower lobes of the lungs is greater than in the upper lobes. The fast upper-chest breathing of people who chronically hyperventilate does not take advantage of the lower parts of the lungs, limiting the amount of oxygen that can be transferred to the blood and resulting in a greater loss of CO_2. Not only this, but upper-chest breathing activates the fight-or-flight response, which raises stress levels and produces even heavier breathing.

Observe your own breathing when you are stressed, or watch the breathing of anxious relatives, friends, or colleagues—you will see that this type of breathing is generally located in the upper chest and goes at a rate that is faster than normal. When we are stressed we tend to overbreathe and resort to breathing through the mouth. Stressed breathing is faster than normal, audible, produces visible movements, and often involves sighs. Many people habitually breathe in this manner every minute of every hour of every day, holding them in a perpetual state of fight-or-flight with adrenaline levels high. The work of even the best stress counselors, psychologists, or psychotherapists will be limited unless they first help their patients to address their dysfunctional breathing. When oxygen delivery to the brain is reduced, no amount of talking and reasoning is going to correct this deficiency. Stressed and anxious patients can only make the progress they really need when their bad breathing habits are addressed.

On the other hand, healthy individuals who are relaxed and relatively free from stress will exhibit breathing that is abdominal: slow, gentle, calm, regular, relatively unnoticeable, silent, and through the nose. To achieve this type of breathing and reduce the negative effects of stressed overbreathing, it is important to activate the body's parasympathetic nervous system to elicit the relaxation response. To do this, you will need to adjust your breathing habits in order to properly use your diaphragm. Avoid sighing, panting, and breathing through the mouth, and become accustomed to slow, gentle, relaxed, calm, and quiet breathing through the nose. This is how we should breathe during rest every minute of every hour of every day. Within a very short time you will find that you feel calmer and more energetic and are able to sleep better. The positive effects of abdominal breathing will continue to transform every aspect of your health, including your sports performance.

Another advantage of abdominal breathing is that it assists with lymphatic drainage. The lymphatic system is effectively the body's sewerage system, draining away waste materials and excess fluid. As the lymphatic system does not have a heart to pump the waste throughout the body, it is reliant on the motions of the muscles, including the diaphragm. During abdominal breathing, lymph is sucked through the bloodstream, neutralizing and destroying dead cells, reducing fluid retention, and improving detoxification of the body.

By utilizing the natural benefits of abdominal breathing you will improve the quality of your blood flow, increase delivery of oxygen to working muscles, and reduce the symptoms of anxiety associated with overbreathing. Returning your breathing habits to the natural and efficient methods you were born with will also allow you to enjoy better health and maximize the potential of your sports or exercise performance. Use the following exercise to encourage abdominal breathing during rest and sports until it again becomes second nature. As the objective of the Oxygen Advantage program is to restore light, abdominal breathing with a high BOLT score, the exercise below forms the foundation upon which remaining exercises are practiced.

Breathe Light to Breathe Right

(*A more advanced version of this exercise can be found on page 255.*)

During the process of breathing, oxygen is drawn into the lungs and excess carbon dioxide is exhaled. The respiratory center located in the brain continuously monitors blood pH, carbon dioxide, and to a lesser extent oxygen. When the level of carbon dioxide in the blood increases above programmed levels, the respiratory center transmits impulses that tell the respiratory muscles to breathe in order to remove the excess gas. When we breathe too much over a period of hours to days, as in the case of chronic stress, the respiratory center adjusts to a lower tolerance of carbon dioxide. Having a lower than normal tolerance to carbon dioxide results in the respiratory center increasing the rate of impulses to the respiratory muscles. The result is habitual overbreathing and excess breathlessness during physical exercise.

You are practicing this exercise correctly when you slow down and reduce your breathing sufficiently to create a tolerable need for air. The need for air signifies an accumulation of arterial carbon dioxide, the goal of which is to reset the respiratory center's tolerance to this gas. To assist with this, it is very helpful to exert gentle pressure against your chest and abdomen with your hands. Try to maintain the need for air for the duration of 4 to 5 minutes.

To practice this exercise, it can be very helpful to sit in front of a mirror to observe and follow your breathing movements.

- Sit up straight. Allow your shoulders to relax. Imagine a piece of string gently holding you up from the top of the back of your head. At the same time, feel the space between your ribs gradually widening.

- Place one hand on your chest and one hand just above your navel.

- Feel your abdomen gently moving outward as you inhale and gently moving inward as you exhale.
- As you breathe, exert gentle pressure with your hands against your abdomen and chest. This should create resistance to your breathing.
- Breathe against your hands, concentrating on making the size of each breath smaller.
- With each breath, take in less air than you would like to. Make the in-breath smaller or shorter.
- Gently slow down and reduce your breathing movements until you feel a tolerable hunger for air.
- Breathe out with a relaxed exhalation. Allow the natural elasticity of your lungs and diaphragm to play their role in each exhalation. Imagine a balloon slowly and gently deflating of its own accord.
- When the in-breath becomes smaller and the out-breath is relaxed, visible breathing movements will be reduced. You may be able to notice this in a mirror.

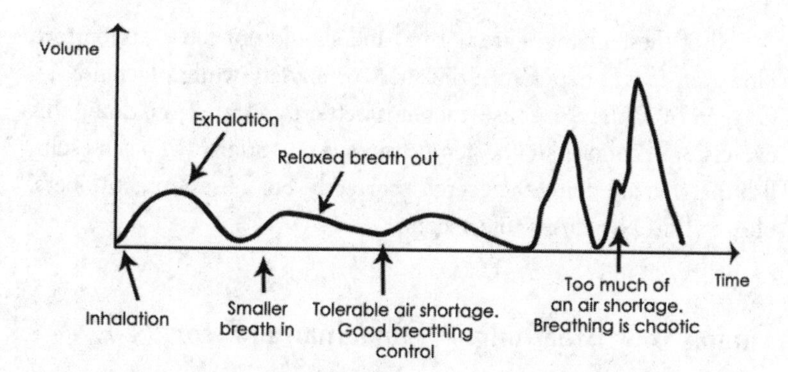

By using a simple exercise like this, you can reduce your breathing movements by 20 to 30 percent. If your stomach muscles start to contract or jerk or feel tense, or if your breathing rhythm becomes disrupted or out of control, then the air shortage is too intense. In this situation, abandon the exercise for 15 seconds or so and return to it when the air shortage has disappeared.

At first, you may only be able to maintain an air shortage for 20 seconds before the urge to breathe is too strong. With practice, you will be able to maintain an air shortage for longer periods. Remember, you are trying to create an air shortage that is tolerable but not stressful. Aim to maintain this tolerable "air hunger" for 3 to 5 minutes at a time. Practicing 2 sets of 5-minute exercises is enough to help you reset your breathing center and improve your body's tolerance for carbon dioxide.

When you practice Breathing Light, the accumulation of carbon dioxide in the blood will result in certain physiological changes in the body. These include:

- A feeling of increased warmth resulting from the dilation of blood vessels
- A rosy red color coming into the face
- Increased production of watery saliva in the mouth, which is an indication that your body is going into relaxation mode and activating the parasympathetic nervous system

All of these changes are normal and should not cause discomfort. However, if you experience dizziness or anxiety while practicing an Oxygen Advantage exercise, then it is better to refrain from doing this exercise and contact an Oxygen Advantage practitioner who can establish that you are doing the exercise correctly. For a list of practitioners, please visit OxygenAdvantage.com.

Timing Your Breathing: A Fundamental Error

You may have noticed that while we attempt to reset breathing volume toward normal, there is no suggestion of changing the number of breaths per minute, or to time the length of each breath. This is deliberate—using time to measure the size of a breath is a fundamental error. Modern Western society seeks to quantify everything in terms of measurement, including our breathing, but when it comes to retraining poor breathing habits, it is not the timing that needs to be focused on. Altering the number of breaths per minute or counting the size of each breath in seconds has taken root in many breathing techniques but actually has little or no effect in addressing poor breathing habits.

For example, telling someone to inhale for 2 seconds and exhale for 3 seconds does not provide guidance on whether they should take in a very gentle breath or a huge inhalation of air. The volume of a gentle breath would be far less than the volume of a large breath, and since we are concerned mainly with volume and reducing it toward normal, counting the length of a breath using seconds does not work. The following illustration shows two different breaths, each with an inhalation length of 2 seconds and an exhalation of 3 seconds. Note the difference in the amount of air taken into the lungs by each breath, despite them being the same length:

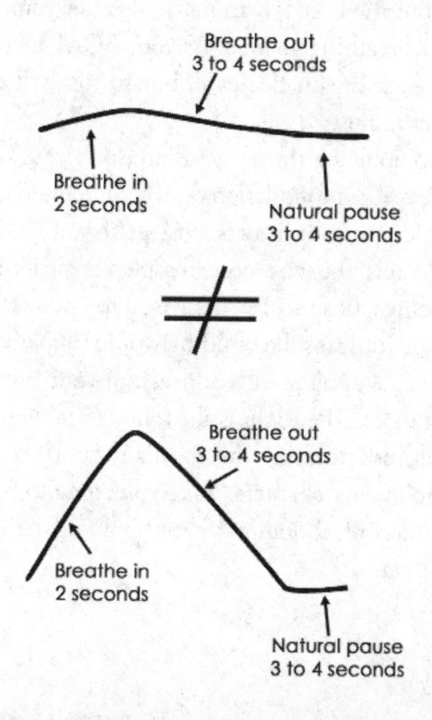

Similarly, is it not possible to address poor breathing habits by changing the number of breaths per minute. Take, for example, an individual who breathes 20 breaths per minute, with each breath consisting of 500ml of air. This breathing pattern provides a total volume of 10 liters per minute. And, since 10 liters of air per minute is too much, the individual may be mistakenly instructed to remedy this by reducing the number of breaths per minute from 20 to 10. However, changing the number of breaths in this way will simply result in each breath doubling in size to compensate for the reduction of breathing rate. There will be no change to breathing volume and the individual's symptoms will persist.

There is only one way to change your breathing volume and rate, and that is by slowing down and diminishing the size of each breath in order to create a shortage of air. In time, as breathing volume changes toward normal, you will obtain a higher BOLT measurement and the number of breaths you take per minute will automatically reduce. To reiterate, it is impossible to change breathing volume by altering breathing rate, but the best way to naturally alter your breathing rate is to reduce your breathing volume. As your BOLT increases, not only will the size of each breath decrease, but so too will the number of breaths taken per minute.

Learning and applying the above exercise to your daily breathing will provide an excellent foundation on which to build a more efficient breathing technique during sports. Just as if you were building the house of your dreams, the most crucial aspect is the foundation. There is no point building a beautiful structure on top of a weak foundation. To do so would be futile, as the building would topple soon after it was built. The same rule applies with addressing your breathing volume. Breathing Light to Breathe Right is the foundation on top of which we simulate high-altitude training during many sports activities, as you will see in the following chapters. Take your time to become familiar with the techniques of abdominal breathing and reduced breathing before you move on.

CHAPTER 5

Secrets of Ancient Tribes

In 1974, twenty-one-year-old Tom Piszkin was attending the University of California at Berkeley. He was a runner and worked part-time in the sporting goods department at Montgomery Ward in Oakland. On October 24 he finished his shift and made his way to a bus stop near the Oakland Coliseum, known as one of the tougher parts of town. Soon after, Tom was confronted by four young men who demanded he give them everything he had. Three of the four youths drew handguns and pressed them against Tom's head, chest, and leg. In shock, when Tom stood up to take his wallet from his pocket, he was shot point-blank in the middle of the chest by a .38 special. The bullet bore through Tom's sternum and entered his left lung. Tom recalls that, surprisingly, the shot didn't cause him a lot of pain.

Following an operation to remove the lead fragments, Tom was discharged from the hospital and resumed running within a month, but recovering from such a traumatic experience was slow and arduous. Tom's journey to restore his lung capacity took over a decade as he continued to fall short of his expectations, despite working hard at his fitness routine. Above all else, Tom wanted to regain the level of athletic fitness he enjoyed before he was shot and was determined to figure out a way to limit his heart rate during his workout sessions. Intuitively, he figured that by reducing the stress on his body during exercise, he would be able to improve his overall fitness and endurance. Tom theorized that it would help to restrict his breathing in order to maintain a steady and manageable exercise intensity—"like a governor on a lawn mower engine"—a solution that was also much cheaper than buying a heart rate monitor.

Tom soon realized that if he couldn't sustain his pace while breathing exclusively through his nose during training, then he was working too hard and going too fast. At first, he found breathing through his nose during physical training a little challenging, but soon discovered that nasal breathing could more easily be maintained by placing tape over his mouth. Not only did Tom tape his mouth during training, he also did it during sleep to ensure that he continued nasal breathing during the night. One year after beginning his reduced breathing training, Tom went for a lung capacity test. His results showed that he was at 130 percent capacity for his weight and age.

Tom has since dedicated his life to his two passions of sports and inventing. He is currently a triathlon coach at the University of California at San Diego and creator of TitanFlex bikes. He is also certified as a USA Olympic Triathlon coach. After thirteen years of serving at the leadership level in the Triathlon Club of San Diego, he was inducted into their hall of fame.

Switching to nasal breathing after spending years mouth breathing requires courage and commitment. Sometimes it is necessary to take one step back to move two forward if you want to truly improve your performance.

If you observe your teammates or competitors during physical exercise, you will no doubt notice that most will be breathing through their mouths. A question I often hear is: "If nasal breathing is so good, then why do most elite athletes breathe through their mouths and not their noses?" A simple answer is that Western breathing habits have become so far removed from what they should be that mouth breathing has become the norm.

Our ancestors used their noses to breathe during physical exercise, as do present-day indigenous tribes, including the Tarahumara, the famed running tribe of Northern Mexico. When researchers studied the nasal-breathing Tarahumara tribe over the course of a 26-mile run, they were astounded to find that their average heart rate was a surprisingly low 130 beats per minute. Compare this with the normal average heart rate of a Western marathon runner, which lies between 160 and 180 beats per minute, and you can see how nasal breathing allows for a calm and sustained breathing pattern even when engaging in intense physical exercise (more on this in chapter 11). Breathing through the mouth is a relatively modern phenomenon and does nothing to improve performance—in fact, it impedes it.

Harvard-educated anthropologist Wade Davis has devoted much of his life's work to the study of indigenous cultures, especially in North and South America. To date, he has lived with fifteen groups of indigenous people, including tribal hunters of the Amazon whose senses are so attuned that they can smell the urine of an animal from forty paces and determine whether it is from a male or female.

While staying with the tribe, Davis, who is also a triathlete, was allowed to accompany them on hunting expeditions. Hunts began in the morning at a pace alternating between a jog and a run. As soon as they picked up the trail of an animal, the hunters switched to running in order to catch up. When the animal became aware of the hunters, it would speed away, but they would continue to follow in close pursuit. The hunters were persistent, increasing their pace so as to give their prey little time to rest, and each time they lost sight of it they would continue on doggedly until they tracked it down again. This pattern could continue over many hours, possibly even days, but the hunters' tenacity would eventually pay off and the animal would collapse from exhaustion, enabling them to capture it at close range. While Davis had a hard time keeping up with the hunters' pace, what was most impressive to him was that they never opened their mouths to breathe. Just like our ancestors, present-day indigenous tribes are able to run at a high intensity with their mouths closed over a relatively long duration of time, a feat that civilized man seems to have forgotten. It is time to go back to basics.

Initially, breathing through the nose might feel strange during exercise, especially if you are used to breathing through your mouth. But remember that the nose is made for breathing, and nasal breathing ensures a number of benefits that are essential not only for good health but for improved sports performance, including:

- Filtering, warming, and humidifying air before it is drawn into the lungs
- Reducing the heart rate
- Bringing nitric oxide into the lungs to open airways and blood vessels
- Better oxygen delivery throughout the body
- Reduced lactic acid as more oxygen is delivered to working muscles

The pace at which you are able to exercise during a half hour of training while breathing through your nose will depend on your BOLT score. Below are some guidelines that are influenced by your nostril and airway size. For example, an athlete with large nostrils will experience less resistance to breathing, enabling exercise at a higher intensity with his or her mouth closed. Here is a general guideline of exercise ability (with the mouth closed) according to BOLT score:

- If your BOLT score is 5 seconds, your ability to walk is seriously hampered. Climbing one flight of stairs will be challenging, requiring you to stop every 3 or 4 steps in order to take a rest.

- If your BOLT score is 10 seconds, you will be able to walk slowly.

- If your BOLT score is 20 seconds, you will be able to walk quickly or jog lightly.

- If your BOLT score is 30 seconds, you will be able to jog at a medium to fast pace.

- If your BOLT score is 40 seconds, you will be able to run quickly.

By increasing your BOLT score, no matter the size of your airways, you will experience lighter breathing and be able to train at a higher pace, for longer distances, while breathing through the nose. Your fitness levels will soon increase well beyond what they were before, and nasal breathing will become much easier during exercise. Within 6 to 8 weeks your BOLT measurement should increase by 10 to 15 seconds, resulting in a significant improvement to your fitness.

Dr. Bill Hang is an orthodontist from California. For the past few decades he has examined the mouths and airways of thousands of patients. What differentiates Dr. Hang from many traditional orthodontists is that when he considers the alignment of the teeth, he also gives attention to the resultant effect on the jaws, the width of the face, and size of airways. In fact, I have yet to meet an individual with such a fascination for airway size.

While we need to be efficient with our use of oxygen, it is also imperative that the size of our airways enables air to flow freely to and from the lungs. If children or teenagers spend five or ten years with their mouths hanging open, their faces will become narrow, their jaws will not develop correctly, and their airway sizes will be reduced. Nasal breathing during the formative years is absolutely essential to help ensure correct development of the face, jaws, and airways. In chapter 13, we will look more closely at the role breathing plays in the development of the face and the need for orthodontic treatment.

I first met Dr. Hang in 2009, when we were both scheduled to give a talk to a study group of myofunctional therapists. As it turned out, we had the same aim in mind: to explore how breathing and resting tongue posture affects sleep, sports, and health. I spoke about the benefits of nasal breathing, while Dr. Hang spoke about airway size and the benefits of having a good facial structure for sports performance. When the airways are too narrow, our ability to perform physical exercise is hampered. Imagine completing a marathon while breathing through airways the size of a very narrow straw—no matter how well trained you are, how fit, how determined, if your airways are constricted you will not be able to take in enough air to properly oxygenate your body.

That day Dr. Hang told me that he had been running for forty-two years and had completed nineteen marathons with his "mouth open like a dog." Following our meeting, he changed to breathing through his nose during training and taped his mouth closed every night to ensure nasal breathing while he slept. Initially he found that his nose would run constantly during exercise, requiring him to blow his nose every few hundred yards. This is a common experience for most people adjusting to nasal breathing as the airways clear and breathing volume increases, but it is a minor inconvenience that will clear up within a few weeks. Like any organ or muscle, the nose needs to get used to being used more during sports.

Six months later, Bill completed the Pasadena Marathon and finished second in his age group. Not only this, but he kept his lips together every step of the way except during a couple of long hills when he allowed his mouth to drop open occasionally. No mean feat for a sixty-year-old! Now, each Sunday, he completes a two-hour run and continues to nasal breathe in order to maintain and ideally improve his fitness. Twenty minutes into his run, when he is good and warm, he is able to get up to a good speed while retaining a slow, regular breathing pattern—much improved from his previous method of panting through the mouth, expending maximum energy as he ran.

Train Your Body to Do More with Less

To reap the most benefit from your physical training, you need to train your body to do more with less. To do this, you will need to reduce your air intake. Incorporating this concept into your training will result in improved breathing economy and an increase in your athletic performance, along with reduced breathlessness and lactic acid during competition. More important, you will not need to push your body beyond its limits, reducing your risk of injury, cardiovascular and respiratory problems, and other health concerns. Nasal breathing during physical training ensures that you do not push yourself beyond what your body is capable of doing.

During physical exercise, there are three ways to reduce air intake:

1. Relax your body and take less air into your lungs.
2. Increase exercise intensity while nasal breathing.
3. Practice breath holding during exercise.

When you first switch to nasal breathing you may find that your ability to train at maximum pace is impeded. Breathing through the nostrils creates resistance and adds an extra load that is likely to slow your performance for the first few weeks. However, with continued practice and by increasing your BOLT score you will soon find that your performance surpasses previous levels.

Competitive athletes who regularly take part in high-intensity training will need to alternate nasal breathing with mouth breathing for an overall improvement to breathing patterns. High-intensity training helps to prevent muscle deconditioning and will require an athlete to periodically breathe through his or her mouth. This is to be expected and can be combined with nasal breathing to attain the best results. For less-than-maximum intensity training, and at all other times, nasal breathing should be employed. For example, competitive athletes may spend 70 percent of their training with the mouth closed, harnessing the benefits of nasal breathing and adding an extra load to their training to increase BOLT score. They may also devote a smaller portion of training to working at an all-out pace in order to maintain muscle condition, for which brief periods of mouth breathing will be required.

During competition there is no need to intentionally take bigger breaths, nor is there a requirement to breathe less. Instead, bring a feeling of relaxation to your body and breathe as you feel necessary. However, breath-holding exercises during your warm-up can be very advantageous, as can practicing breathing recovery during your cooldown. Competition isn't the ideal time to focus on how well or how poorly you are breathing, as your full concentration should be devoted to the game or race. The best way to improve breathing for competition is to improve your everyday breathing, and the key to this is obtaining a higher BOLT score.

Recreational athletes who are not taking part in competition or high-intensity exercise, however, are far better off maintaining nasal breathing at all times. While reducing your breathing during physical exercise, try not to overdo it. If you find that your need for air is so great that you need to open your mouth, simply slow down and allow your breathing to calm once more.

Unveiling the Warm-Up

The vast majority of sports coaches are in agreement that warming up before exercise is vitally important. During physical exercise the body requires greater blood flow to tissues and muscles than during rest. The objective of a warm-up is to increase blood flow and prepare the body for more intense physical exercise, thus reducing the incidence of sports-related injuries and improving overall performance.

It takes time for the body to warm up, but when it does, your body is able to function more effectively during exercise. When your body is warmed up prior to exercise, the following benefits can be maximized:

- The production of more carbon dioxide—improving the release of oxygen from the blood to tissues and organs—increasing VO_2 max, improving endurance, and reducing the risk of injury
- The opening of blood vessels and airways—allowing for better blood flow and easy breathing

In practice, however, many athletes do not warm up sufficiently. Many spend, at most, 2 to 3 minutes warming up by doing a light jog before intensifying their exercise. This is simply a case of going too fast, too soon.

Aisling plays soccer for one of the better amateur football teams in

Ireland. She has reached a high level of fitness and is a newcomer to the Oxygen Advantage exercises. Despite her excellent fitness, she often finds that she feels quite breathless during the first 10 to 20 minutes of a game, while at the end of the game she feels as if she could go on playing forever. This is a common complaint within the world of sport, and usually comes down to a lack of warm-up. The best way to avoid early breathlessness during exercise is to increase your BOLT score and spend more time warming up with nasal breathing.

For people like Aisling, who find it hard to get going when they exercise, at least 10 minutes should be spent warming up before physical activity, especially during cold weather, as it can take the body up to 30 minutes to function at its best. In order to devote all your energy to a game, your body should reach peak form early on rather than somewhere during the second half. By skipping an adequate warm-up—either through impatience or because you believe it's not necessary—you are cheating yourself out of performing at your full capability.

To maximize the benefits of your warm-up, combine movement with relaxation techniques and breath holds, as described below.

Oxygen Advantage Warm-Up

- Begin walking at a pace that is comfortable for you.
- During your warm-up, try to breathe regularly and calmly through your nose, using your diaphragm to maintain a gentle and relaxed breathing technique.
- Feel your abdomen gently moving outward as you inhale and gently moving inward as you exhale.
- As you walk, allow a feeling of relaxation to spread throughout your body. Silently encourage the area around your chest and abdomen to relax (you will find that any tension can be released by silently telling that

area of the body to relax). Feel your body relax and become soft. Body relaxation during physical exercise helps to ensure steady, calm, and regular breathing.

- After 1 minute or so of walking at a fairly good pace, exhale normally through your nose and pinch your nose with your fingers to hold the breath. (If you are in a public place, you might prefer to hold the breath without holding your nose.)
- While holding your breath, walk for 10 to 30 paces, or until you feel a moderate need to breathe. When you feel this hunger for air, let go of your nose and resume breathing through your nose.
- Continue walking for 10 minutes, performing a breath hold every minute or so.

Creating an air shortage by holding the breath during your warm-up is vitally important to cause an accumulation of carbon dioxide in the blood before physical exercise commences.

Your breathing will naturally increase when you exercise more intensely, but without a corresponding increase in the production of CO_2, the result will be a net loss of carbon dioxide. This loss can lead to reduced oxygen delivery to working muscles and the constriction of airways and blood vessels. Not surprisingly, most asthma attacks and breathing difficulties occur during the first 10 minutes of physical exercise.

In order to avoid exercise-induced asthma, there are three simple guidelines:

1. Attain a high BOLT score
2. Nasal breathe
3. Warm up

Breathe Light to Breathe Right—Jogging, Running, or Any Other Activity

After a 10-minute warm-up of walking using relaxation and breath holds, you can increase your pace to a jog or run. Start off easy and follow your breathing, continuing to breathe through the nose. While jogging or running, it is necessary to keep your breathing regular and under control. If you find it too difficult to run with your mouth closed, it simply means that your pace is too fast at that moment. Slow down or even walk a little until you recover, always breathing in and out through the nose, especially if your BOLT score is less than 20 seconds.

You can check whether you are pushing yourself too hard during physical exercise by exhaling normally and holding your breath for 5 seconds. When you resume breathing through the nose, your breathing should remain controlled. If you find that you lose control of your breathing, you are pushing yourself too hard.

No matter what type of exercise you prefer, make sure that you watch your breath and become aware of your inner body. Silently repeat the word *relax* to help release the tension around your abdomen. Bring your entire attention from your mind into your body. Become one with your run or exercise, merging body, mind, and activity together. Move with every cell of your body, from the top of your head to the tips of your toes. Doing so will give your training, your sport, and your competition the attention they deserve. When engaged in an activity, there is no point to just going through the motions. One needs to become the activity.

As you run, feel each gentle connection between your feet and the ground as you propel yourself forward. Avoid pounding the pavement as this will lead to sore hips, sore joints, and possible injuries. Instead, bring a feeling of lightness to your body, and visualize yourself barely touching the ground as your run. Imagine yourself running over thin twigs, treading so softly that they do not break, emulating the words of Chinese philosopher Lao Tzu: "A good runner leaves no tracks." Remember: light foot strikes, a relaxed body, and regular, steady breathing.

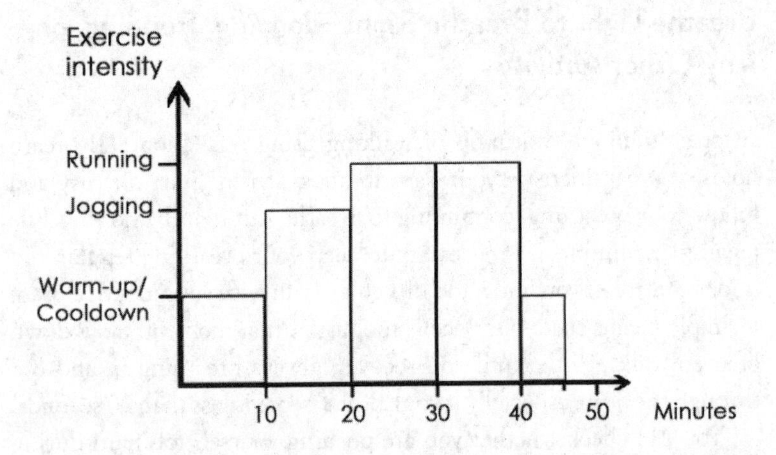

Ten or fifteen minutes into your run or other workout, you will probably experience an endorphin high from your exercise. Allow your body to find its perfect work rate by breathing through your nose in a steady and regular fashion. There is no need whatsoever for a heart rate monitor to provide feedback on training intensity. Instead, let your nose, your breathing rhythm, and how you feel dictate your training intensity. Continue to increase your pace to the point that you can maintain steady and regular breathing through the nose. If your breathing rhythm becomes chaotic or it is necessary to open your mouth to breathe, your exercise intensity is too much. If this happens, slow your pace to a walk for 2 to 3 minutes to allow your breathing to recover. When you are able to breathe calmly through your nose once more, you can resume your physical exercise.

As you continue your physical exercise, the increased generation of carbon dioxide and heat will enhance the delivery of oxygen from the blood to working muscles, as well as facilitating the dilation of airways and blood vessels. Your body will be warm and sweating, your breathing will be faster than normal but steady, and your head will be clear. If you keep your mouth closed throughout your exercise, your breathing will recover quickly.

Breathing Recovery Exercise

Following physical exercise, cool down by walking for 3 to 5 minutes, performing the following small breath holds:

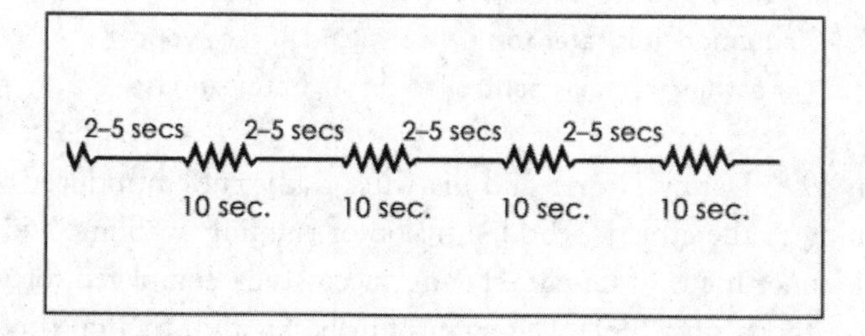

- Exhale as normal through the nose.
- Pinch your nose with your fingers to hold the breath for 2 to 5 seconds.
- Breathe normally through the nose for 10 seconds.
- Repeat the first 3 steps throughout your cooldown.
- Resume regular breathing.

How to Determine Whether You Have Exercised Correctly

In addition to using breath holds to determine your training limits, you can also use your BOLT score to see how efficient your breathing is during physical exercise. Follow these steps to track your progress using your BOLT score:

- Measure your BOLT score before training.
- Perform your physical exercise.
- Measure your BOLT score one hour after you finish training.
- If your BOLT score is higher after exercise than before, your breathing is efficient during exercise.
- If your BOLT score is lower after exercise than before, your breathing is inefficient during exercise. In this situation, it is safer to slow down and ensure your breathing remains controlled throughout exercise.

In 1999 Danny Dreyer and his wife, Katherine, introduced Chi-Running to the athletic world, a fusion of running, walking, and the subtle inner focus of tai chi. Having successfully completed forty ultramarathons since 1995, Danny has finished within the top three of his age group in all but one competition. A firm exponent of nasal breathing, Danny advises the use of nose breathing as a self-regulating mechanism, "because you won't be able to do it if you are either running too fast, not relaxed enough, or inefficient in your movement." Similar to the experience of most seasoned runners, when Danny first began nose breathing he could only go for a minute or so before he had to breathe through his mouth. However, as he became more efficient and his breathing relaxed into his running form, he was able to breathe through his nose for longer periods of time. Another reason for breathing through the nose, according to Danny, is that it allows air more fully to be brought into the deeper parts of the lungs, enabling a better gas exchange to take place.

See for yourself. Switch to breathing through your nose and you will notice the benefits within a short space of time.

PART II

The Secret of Fitness

CHAPTER 6

Gaining the Edge—Naturally

According to the United States Olympic Training Center, there is a difference of less than 0.5 percent in the performance of Olympic athletes. With such a narrow margin for success, it is essential for athletes and coaches to look for new ways to attain a competitive edge. Since oxygen is the fuel for working muscles, anything that increases body oxygenation above normal levels will be of great benefit to an athlete's performance, and because oxygen is a natural and limitless resource, it is the ideal tool for boosting performance legally.

One way of tapping into your own natural resources is to purposefully subject the body to reduced oxygen intake for a short period of time. When the human body is exposed to situations in which there are reduced oxygen levels—such as high altitude, or by holding the breath—adaptations take place that force the body to increase oxygenation of the blood. Even if you are not a competitive athlete, using these techniques will allow you to get the most out of your workout and accelerate whatever fitness program you undertake. Who doesn't want to do more with less effort?

While striving for improved performance, however, there will always be athletes who choose to participate in illegal methods of blood doping, either through blood transfusion or by taking a banned substance such as erythropoietin (EPO), testosterone, or human growth hormone.

Blood transfusions are a testament to the drastic and illegal measures that some athletes go to in order to gain the edge over their competitors. A few weeks before competition, blood is extracted from the athlete's body and stored in a freezer or refrigerator. The body, sensing that blood levels are lower than normal, will produce additional red blood cells to bridge the gap. Closer to the endurance event, typically between one and seven days, the stored blood is reinfused into the body of the athlete. This addition of blood increases quantities of red blood cells above normal levels, which in turn increases VO_2 max and enhances physical performance.

By the early 1990s EPO had become the banned substance of choice amongst athletes seeking to increase their endurance performance. EPO is a naturally occurring hormone produced in the kidneys that stimulates the bone marrow to release more red blood cells into circulation. Because red blood cells carry oxygen from the lungs to the muscles, having a higher concentration in the circulation can greatly improve an athlete's aerobic capacity. EPO that is produced in a lab is almost identical to the naturally occurring hormone that is produced in the body. For medical purposes, EPO is prescribed to people with chronic kidney disease–induced anemia as their condition results in a decrease in the amount of red blood cells in circulation. However, soon after its inception, some members of the athletic community realized that taking an artificial version of EPO improved sports performance by increasing the oxygen-carrying capacity of the body.

The most notorious endurance event on earth for blood doping has to be the Tour de France. Regarded as the most prestigious of all cycle races in the world, and with places limited to around two hundred athletes, participation in the Tour de France is the dream of any budding amateur or professional cyclist. Not for the fainthearted, the event consists of a grueling 2,200-mile cycle over twenty-two days, with some mountain climbs lasting for twenty miles or more. Ever since its inception in 1903, there have been allegations of cyclists resorting to various illegal techniques to help them complete the event or improve their performance. Early reports involved cyclists fueling their bodies with alcohol, stopping off at various stages to load their pouches with wine, beer, or whatever else they could get their hands on, more so to numb the pain to help finish the race rather than improve athletic performance. However, in more recent decades, competitors took greater risks to gain an edge.

A granite memorial to the renowned British cyclist Tom Simpson stands on the spot where he collapsed and died during the 1967 Tour de France. The epitaph reads: OLYMPIC MEDALIST, WORLD CHAMPION, BRITISH SPORTING AMBASSADOR. At twenty-nine, Simpson was regarded as one of the all-time best British cyclists. During the race, as the route made its way through the Alps, Simpson fell ill with diarrhea and stomach pains.

In the searing heat, close to the summit of Mont Ventoux, Simpson fell. With a determined effort to continue, he ordered onlookers to "put me back on my bike," and continued to ride 500 yards farther before collapsing again. Despite efforts by a nurse to resuscitate him, he was pronounced dead after being airlifted by helicopter to the hospital. Simpson's postmortem revealed amphetamines in his system. Later, investigators would discover more evidence of the drugs in his hotel room and the pockets of his jersey.

In later years, methods of doping became more sophisticated. Tyler Hamilton, who was the now-disgraced former champion Lance Armstrong's teammate, described how he got goose bumps as blood, fresh from the refrigerator, entered his veins. In his book, *The Secret Race,* Hamilton claims that Armstrong also had a blood transfusion to improve his performance, and that during the 1998 Tour de France the riders were followed on a motorbike by an accomplice carrying fresh vials of EPO. "To Lance's way of thinking, doping is a fact of life, like oxygen or gravity," Hamilton wrote.

Lance Armstrong's fall from grace came on October 10, 2010, when the U.S. Anti-Doping Agency (USADA) released a statement concluding that the "evidence shows beyond any doubt that the U.S. Postal Service Pro Cycling Team [Lance Armstrong's team] ran the most sophisticated, professionalized, and successful doping program that sport has ever seen." Summed up in the statement was the courage of eleven of Armstrong's former teammates who participated in the doping conspiracy but assisted the agency in its investigation in order to "help young athletes have hope that they are not put in the position they were."

In January 2013, in a no-holds-barred interview with Oprah Winfrey, Armstrong admitted to taking banned substances, including EPO, testosterone, human growth hormone, and cortisone, and confessed to blood doping and blood transfusions to enhance his cycling performance. When Winfrey asked if he had used illegal substances or doping methods during all seven of his Tour de France victories, the crushing answer was "yes."

Preparation to compete in the Tour de France often takes place at a young age. From their early teens, cyclists sacrifice their social lives and spare time to cycling and training to build strength, stamina, and endurance. I would like you to put yourself in this situation for a moment. Imagine that for years you have devoted every waking hour to training, living, and dreaming cycling. After a roller-coaster few years, you are finally good enough to take part in your greatest aspiration: the Tour de France. But soon within your first season, your colleagues present you with two options: Either blood dope and have some chance of competing on a level playing field, or choose not to blood dope and return home, leaving your dreams behind. This is likely the scenario that faced many cycling greats, including Tyler Hamilton, Floyd Landis, Bjarne Riis, and Marco Pantani, who wanted nothing more than to compete in the sport they loved. While many cyclists reluctantly gave into the temptation, others chose to abandon their chance at the Tour de France. Stephen Swart grew up on New Zealand's North Island, and in his junior racing days both he and his brother were very successful cyclists. Swart cycled alongside Lance Armstrong in 1994 and 1995, but at age thirty he walked away from cycling completely and was later vilified by his fellow cyclists for "spitting in the soup" after he broke the code of silence about doping within the sport. Looking back, Swart said he felt cheated in a way, wishing he had never been put in the position to dope. His natural ability was undermined by the doping culture that surrounded him. For many years, winning the Tour de France seemed to be as much about whose doctor prescribed the best cocktail of banned substances than the athletic prowess of the competitors.

Since the sporting world has come under increased attention from investigative journalists—including David Walsh, chief sportswriter with the *London Sunday Times*, and Paul Kimmage, former professional cyclist and award-winning sports journalist—dealing with cheating has now risen to the top of the agenda for many sporting authorities. Kimmage, who spent the past few decades exposing the doping culture in the Tour de France, commented: "I've always understood the pressure to dope. I've always understood the temptation to dope, and I understand because I've been there. The perception of the Tour de France now from the public is that it's rotten, they all dope, and that saddens me because it shouldn't have happened."

Fortunately for the future of sports, the culture is slowly changing, and the majority of athletes do not partake in the unethical practice of blood doping. Instead, they choose naturally beneficial activities such as high-altitude training or other techniques designed to increase the body's ability to carry more oxygen.

The main purpose of altitude training and the Oxygen Advantage techniques outlined in this book is to increase red blood cell count. By practicing the breath-hold exercises outlined in this book, the kidneys increase production of EPO and the spleen releases red blood cells into the blood circulation. Both of these effects increase the oxygen-carrying capacity of the blood above normal levels, giving an athlete a competitive edge without the risks and ethical issues of illegal doping. A higher concentration of red blood cells can benefit your sporting performance in several ways, including:

- Improving the oxygen-carrying capacity of your blood
- Increasing your VO_2 max
- Extending your endurance potential

Maximal oxygen uptake, or VO_2 max, refers to the maximum capacity of an individual's body to transport and utilize oxygen during 1 minute of exhaustive exercise. The V refers to volume, the O_2 to oxygen, and max to the maximum capacity of your body. Your VO_2 max is measured by the amount of oxygen that is used during 1 minute of exercise per kilogram of body weight. VO_2 max is a factor that can determine an athlete's capacity to sustain physical exercise, and is considered to be the best indicator of cardiorespiratory endurance and aerobic fitness. In sports that require exceptional endurance, such as cycling, rowing, swimming, and running, world-class athletes typically have a high VO_2 max. The goal of most endurance programs is to increase an individual's VO_2 max, and this can be achieved by improving the oxygen-carrying capacity of the blood.

The rest of this chapter explores several different training regimens along with their effects on VO_2 max and the oxygen-carrying capacity of the blood. In order to understand how and why these techniques work, it is useful to know the following basic information about the composition of your blood and some common terms that we will be referring to regularly.

Blood is made up of three parts: oxygen-carrying red cells, white blood cells, and plasma. *Hemoglobin* is a protein found within the red cells. One of the functions of hemoglobin is to carry oxygen from the lungs to the cells, tissues, and organs of the body, where it is released in order to burn nutrients for the production of energy. Once oxygen has been released, the resultant carbon dioxide is collected by hemoglobin and returned to the lungs, which exhale the excess.

Levels of hemoglobin will vary from person to person, but the following figures provide a general guide for normal results:

Male: 13.8 to 17.2 gm/dL

Female: 12.1 to 15.1 gm/dL

(gm/dL = grams per deciliter)

Hematocrit refers to the percentage of red blood cells in the blood. Under normal conditions, hematocrit will relate closely to the concentration of hemoglobin in the blood. Hematocrit is usually found to be 40.7 to 50.3 percent for males and 36.1 to 44.3 percent for females.

Another measurement relevant to Oxygen Advantage techniques is the oxygen saturation percentage of hemoglobin. Hemoglobin has a maximum oxygen-carrying capacity, and oxygen saturation simply means how much of that capacity is filled with oxygen. Normal arterial oxygen saturation is between 95 and 99 percent.

In the following sections we look at research that investigates supplementary training programs, including high-altitude training, high-intensity exercise, and simulation of high altitude by breath holding, and compare how these techniques can improve oxygen-carrying capacity and athletic performance naturally.

The Merits of High-Altitude Training

Traditional altitude training methods involve living and training at a high altitude, forcing the body to adapt to exercising with less oxygen and therefore increasing the blood's oxygen-carrying capacity. Athletes still use this technique today, particularly those who live at high altitudes such as Kenyan and Ethiopian runners. However, there is a significant drawback to training at high altitude, since exercising in such an atmosphere increases resistance, which can prevent an athlete from achieving his or her maximum work rate. This reduction in exercise intensity can result in muscle deconditioning.

To limit the detraining effects of working at high altitude while still maintaining the benefits, Dr. Benjamin Levine and Dr. James Stray-Gundersen from the University of Texas in Dallas developed the "live high and train low" model in the 1990s. This model requires an athlete to live at a moderate altitude of 2,500 meters but to train at an altitude lower than 1,500 meters. The premise of the method is to enable athletes to benefit from the positive physiological changes associated with living at a high altitude while enabling them to train at their maximum work rate.

Levine and Stray-Gundersen conducted a study of thirty-nine male and female collegiate distance runners who were evenly matched in fitness level. Each runner was assigned to one of three groups:

1. Live low (150 meters) and train low (150 meters)
2. Live high (2,500 meters) and train low (1,250 meters)
3. Live high (2,500 meters) and train high (2,500 meters)

Results for the second group, "live high and train low," showed a 9 percent improvement in red blood cell volume and a 5 percent improvement in maximal oxygen uptake (VO_2 max). The improvement in maximal oxygen uptake was in direct proportion to increased red cell mass volume. This translated to an impressive performance improvement of 13.4 seconds in a 5,000-meter run.

After returning to sea level, the "live high and train low" group was the only one to demonstrate significant improvements in both VO_2 max and 5,000-meter run time. These improvements were attributed to the athletes' acclimatization to altitude while maintaining the velocity of their sea-level training, most likely also accounting for the increase in their VO_2 max.

Another study replicated these results using national team distance runners. Following 27 days of training at an altitude of 2,500 meters, participants achieved an improvement of 1.1 percent in a 3,000-meter time trial. Although a 1.1 percent improvement in performance may not seem to be a large effect, at an elite level, races are won or lost by small fractions of a percent. Furthermore, the increase in running performance was accompanied by a 3 percent improvement in maximal oxygen uptake.

The United States national team for long track speed skating utilized the "live high, train low" model to prepare for the 2002 Winter Olympics in Salt Lake City. That year they enjoyed unprecedented success, with six athletes winning eight medals (three of which were gold) and two world records broken. During the 2006 Torino Olympics, the U.S. long track speed skaters who continued to employ the "live high, train low" model brought home three gold, three silver, and one bronze medal.

The Merits of High-Intensity Training

Another training method that receives considerable attention from athletes and coaches is high-intensity training. The fundamental principle of high-intensity training is to exercise in short, intense bursts, performing at maximum work rate—a technique that is certainly not for the fainthearted. Numerous studies have investigated the different responses from training at different intensities, and in comparison to moderate exercise, high-intensity training provides greater improvements to both aerobic and anaerobic capacity. Aerobic exercise is related to endurance and ensures that the body is supplied with enough oxygen to continue to perform. Anaerobic exercise means "without oxygen" and is more concerned with speed, power, and strength, leading to improved performance in a shorter space of time.

Japanese scientist Izumi Tabata and colleagues at the National Institute of Fitness and Sports in Japan conducted a study of two training experiments to compare moderate- to high-intensity training. The high-intensity group participated in the method known as Tabata training, in which athletes give their full effort at an exhausting work rate for periods of just 20 seconds at a time. The authors of the study concluded that although moderate-intensity aerobic training improved aerobic power, high-intensity intermittent training improved both anaerobic and aerobic performance.

In another study, Stephen Bailey and colleagues from the University of Exeter in the UK compared a high-intensity sprint training program with low-intensity endurance training, measuring VO_2 uptake and muscle deoxygenation. Posttrial results showed that the high-intensity group experienced faster VO_2 kinetics and an increased tolerance of high-intensity exercise. This means that the athletes experienced faster oxygen uptake when transitioning between rest and exercise, allowing them to perform at a higher standard more easily. This improved oxygenation of active muscles also contributes to decreased recovery time following exercise and a reduction in the production of lactic acid.

It seems clear, therefore, that high-intensity training offers several positive benefits to athletes, including:

- Improved anaerobic and aerobic energy supplying systems, allowing for greater endurance, strength, speed, and power
- Faster VO_2 kinetics, allowing the blood to carry more oxygen to the muscles
- Increased tolerance to high-intensity exercise
- Decreased recovery time from less than maximum exercise
- Reduced lactic acid buildup
- Improved oxygenation of active muscles, allowing you to exercise harder and longer

In the next section we will examine how to produce the beneficial effects of high-altitude and high-intensity training to increase exercise performance.

The Science of Simulating High-Altitude and High-Intensity Exercise

High-altitude training in real conditions is obviously more feasible for athletes living in countries such as Kenya than those of us living in, say, Ireland, where the low-lying terrain reaches no higher than 1,000 meters. Similarly, high-intensity training may not be practical for some people, as it involves maximum physical effort and respiration until exhaustion. Some people will find high-intensity exercise extremely uncomfortable or find that they lose control of their breathing, which can lead to health concerns.

A practical alternative available to all athletes regardless of location and fitness is to supplement regular training with breath-hold training. In the following sections we will learn how breath-holding techniques allow us to simulate many of the positive benefits of high-altitude and high-intensity training, including:

- The release of red blood cells from the spleen, improving aerobic performance
- The production of natural EPO
- A higher tolerance to carbon dioxide
- Reduced stress and fatigue of working muscles
- Improved psychological preparedness
- Improved recovery time
- Reduced lactic acid
- Improved swimming technique (as discussed later)
- The ability to maintain fitness during rest or injury
- Maintenance of these benefits without the need to travel to high altitudes

For hundreds of thousands of years, breath holding was practiced extensively by our ancestors for the purposes of foraging for food by diving in a deep-water environment, and some evolutionary theorists even suggest that it might have been responsible for a number of unique human features. To this day, predominantly female Japanese pearl divers known as *ama* continue with the tradition of breath-hold diving—a practice thought to be over two thousand years old.

The best-equipped natural diver is likely the Weddell seal, which can remain submerged in water for up to two hours at a time. Although humans do not have the same adaptive physiological response of the seal, we are able to exhibit certain coping mechanisms in order to deal with a relative lack of oxygen. Generally, most humans can hold their breath after an inhalation for a maximum of up to about 50 seconds, with elite divers achieving a static breath hold of between 8 minutes 23 seconds and 11 minutes 35 seconds.

A number of studies have sought to understand the significant role that breath holding can play in adapting the body for increased oxygen delivery, with researchers investigating the effects of breath-hold diving in native divers, professional divers, and untrained divers.

The spleen is an organ that acts as a blood bank; when the body signals an increased demand for oxygen, the spleen releases stores of red blood cells. It therefore plays a very important role in regulating blood hematocrit (the percentage of red blood cells in the blood), as well as hemoglobin concentration.

Provoking the body to release additional red blood cells and increase the concentration of hemoglobin in the blood improves the body's ability to deliver oxygen to working muscles during exercise. Breath-holding studies involving volunteers whose spleens had been removed for medical reasons demonstrate just how vital this organ is in changing the composition of the blood. After a series of short breath-holding exercises, those with spleens intact showed an increase in hematocrit and hemoglobin concentration of 6.4 percent and 3.3 percent respectively, while those without spleens showed no alterations in blood composition at all. This means that after as few as 5 breath holds, the oxygen-carrying capacity of the blood can be significantly improved with the help of the spleen.

This organ also influences how long a person can hold their breath for. In one study, participants were able to achieve their longest breath hold on their third attempt. Trained breath-hold divers peaked at a total of 143 seconds, untrained divers at 127 seconds, and splenectomized volunteers—those who had previously had their spleens removed—achieved 74 seconds. Not only that but spleen size decreased by a total of 20 percent in both breath-hold divers and the untrained volunteers, demonstrating a rapid contraction of the spleen in response to the reduction of oxygen. What this means is that breath-holding ability improves with repetition as the spleen contracts, releasing additional red blood cells into circulation and improving the oxygen-carrying capacity of the body. While these studies generally include subjects holding their breath for as long as possible, significant splenic contraction has been found to take place with even very short breath holds of 30 seconds. However, the strongest contractions of the spleen, and therefore the greatest changes to blood composition, are shown following maximum breath holds.

Another useful piece of information gleaned from these studies is that it is not necessary to be immersed in water to benefit from the effects of breath-hold diving. There seems to be no discernible difference between the increase of hematocrit and hemoglobin concentration in volunteers practicing breath holds in and out of water. Since there is no visible increase in the results of breath holding with the face immersed in water, it can be concluded that it is the breath hold itself that stimulates splenic contraction. In other words, it is not being underwater that causes the spleen to release red blood cells into the circulation but the simple drop of oxygen pressure in the blood resulting from holding of the breath. Therefore, the benefits of breath holding are not limited to divers and swimmers. This is of particular relevance to the Oxygen Advantage program, as our breath-hold exercises are performed out of water.

The relevance of the above studies suggests that effects similar to those achieved with high-altitude training can be obtained at sea level simply by performing a series of breath holds. Stimulating the spleen to contract by reducing the availability of oxygen causes an increase in hemoglobin and hematocrit, which in turn increases the oxygen-carrying capacity of the blood and improves aerobic ability.

The most appealing aspect of breath holding is that it is feasible for most individuals and is not as taxing on the body as high-intensity exercise. Performing just 3 to 5 breath holds of maximum duration can lead to a 2 to 4 percent increase in hemoglobin. This might not sound like much, but where a fraction of a second can determine the difference between the winner and the loser, every possible advantage counts.

Why Oxygen Advantage Training Elicits an Even Stronger Response

In the studies investigating breath holds and splenic contractions discussed, each breath hold was performed following an inhalation. You might be wondering why Oxygen Advantage breath holds are performed after an exhalation. Let me explain.

Performing a breath hold after an exhalation lowers the oxygen saturation of the blood to simulate the effects of high-altitude training. I have monitored the blood oxygen saturation of thousands of individuals as they practice breath holds, and by far the greatest change to oxygen saturation occurs after an exhalation. For most people, after four or five days of practice, a drop of oxygen saturation below 94 percent can be observed—a level comparative to the effects of living at an altitude of 2,500 to 4,000 meters.

Gently exhaling prior to holding the breath reduces air content in the lungs, allowing a quicker buildup of carbon dioxide and eliciting a stronger response. While this reduces the length of time for which you can hold your breath, increased carbon dioxide has been shown to improve hemoglobin concentration by around 10 percent compared to a breath hold with normal carbon dioxide.

Higher levels of carbon dioxide in the blood can produce an even greater contraction of the spleen, resulting in an increase in the release of red blood cells and therefore the oxygenation of the blood.

Increased CO_2 in the blood also causes a rightward shift of the oxyhemoglobin dissociation curve. As described by the Bohr Effect, an increase in carbon dioxide decreases blood pH and causes oxygen to be offloaded from hemoglobin to the tissues, further reducing blood oxygen saturation.

Holding the breath on the exhale also capitalizes on the benefits of nitric oxide by carrying the gas into the lungs rather than expelling it. By exhaling and holding the breath, nitric oxide is able to pool in the nasal cavity so that when breathing resumes, air laden with nitric oxide is inhaled into the lungs.

Increase Erythropoietin (EPO) Naturally

As we have seen, erythropoietin, often known as EPO, is a hormone secreted by the kidneys in response to reduced oxygen levels in the blood. One of the functions of EPO is stimulating the maturation of red blood cells in the bone marrow, thereby increasing oxygen delivery to muscles. Breath holding is an effective way of stimulating the release of EPO, allowing you to fuel your blood with increased levels of oxygen and enhance your sports performance. The concentration of EPO can increase by as much as 24 percent when the body is subjected to lower oxygen levels using breath-hold exercises.

A clear example of the relationship between breath holding and EPO production can be found in those suffering from sleep apnea. Sleep apnea is a condition involving involuntary holding of the breath after exhalation during sleep. Depending on the severity, the sleeper may hold their breath from 10 to 80 seconds, and this may occur up to 70 times an hour. During sleep apnea, the oxygen saturation of the blood with oxygen can reduce from normal levels of around 98 percent to as low as 50 percent. These reduced oxygen levels can cause an increase in EPO of 20 percent.

Of course, there is quite a difference between the condition of sleep apnea and the practice of breath holding to enhance sports performance. However, it is interesting to note the effect of breath holding (both voluntary and involuntary) on the production of natural EPO. Increasing EPO levels allows the blood to deliver greater amounts of oxygen to the muscles and is the natural equivalent of the illegal blood doping methods discussed at the beginning of this chapter. The benefit of using breath holding as a performance-enhancing exercise is that, unlike sleep apnea, conscious breath holding allows you to keep complete control over the frequency and duration of each hold. And, unlike blood doping, the EPO you produce using simple breathing techniques is free, effective, and legal.

The Importance of Movement for Simulating High-Altitude Training

During physical exercise or breath holding, a shortage of air is created. An air shortage is best described as a hunger or desire to breathe, varying in intensity from mild to moderate to strong. The intensity of air shortage will differ depending on the exercise or situation. For example, while practicing the exercises in this book, the air shortage during sitting should be mild or tolerable whereas the air shortage during intense physical exercise can be strong. A strong air shortage during physical exercise is beneficial during training as it conditions the body to tolerate extreme demands, and is often popular among athletes as it presents a new challenge to pit their willpower and determination against.

A strong air shortage during physical exercise is more suitable for athletes with a BOLT score of longer than 20 seconds. When your BOLT score is shorter than 20 seconds, you must be careful not to hold the breath for too long as it can cause a loss of control of your breathing. It is important that you are always able to resume calm breathing following a breath hold. The shorter the BOLT score, the easier it is to lose control of your breathing.

Please note that when creating an intense air shortage you may develop a headache as your blood oxygen saturation decreases, but this should disappear after about 10 minutes of rest. Try to avoid overdoing the exercises to the point that they bring on a headache.

Breath Holding to Improve Respiratory Muscle Strength

The respiratory center is located in the brainstem and continuously monitors blood oxygen, carbon dioxide, and blood pH, using this information to control the amount of air taken into the body. When the body requires a breath of fresh air, the brain sends a message to the respiratory muscles, telling them to breathe. The diaphragm, which is the main breathing muscle, moves downward, creating negative pressure in the chest cavity, resulting in an inhalation. Following inhalation, another message is sent for the diaphragm to move back to its resting position, causing an exhalation.

When the breath is held following an exhalation, the intake of oxygen is halted while carbon dioxide accumulates in the blood. During this pause, oxygen cannot enter the lungs, and carbon dioxide cannot leave the bloodstream. The respiratory center, noticing the change to blood gases, communicates to the diaphragm to resume breathing, and the diaphragm contracts downward in an attempt to allow the body to breathe. However, breathing cannot resume while the breath is held, and the brain begins to send increasingly frequent messages to the diaphragm, causing its spasms to intensify. You can experience this by simply holding your breath until you feel a strong need to breathe. At first you will feel an isolated spasm of the diaphragm, but this will soon be followed by more intense and quicker spasms as the body attempts to resume breathing.

In essence, holding the breath until a medium to strong need for air mobilizes the diaphragm, provides it with a workout and helps to strengthen it. While there are many products on the market aimed at increasing respiratory muscle strength, breath holding may be the easiest and most natural as it can be employed at any time and actively directs attention to the diaphragm. Improving respiratory muscle strength can be extremely beneficial during exercise, especially when fatigue of the diaphragm may determine exercise tolerance and endurance.

Breath Holding to Reduce Lactic Acid

Just as injury plays a role in limiting physical performance, mental and physical fatigue can also prevent an athlete from pushing harder and faster. As U.S. Army general George Patton wrote to his troops during World War II: "Fatigue makes cowards of us all. Men in condition do not tire." And he was right; endurance is relative to how well the body is prepared, and the onset of fatigue occurs when the body is pushed beyond the limits of preparation.

Working a muscle without sufficient fuel generates lactic acid, and while small amounts can be beneficial, acting as a temporary energy source, a buildup of lactic acid creates a burning or cramping sensation in the muscle that can slow down or even halt exercise completely.

Studies with athletes have demonstrated that breath holding after an exhalation deliberately exposes the body to higher levels of acidity, thereby improving tolerance and delaying the onset of fatigue during competition.

In a team sport like football, where players are expected to maintain form and concentration during 90 minutes of intense activity, the ability to push through or avoid fatigue can be instrumental to a team's success. I recently worked with the Galway women's football team, whose coach, Don O'Riordan, was concerned that the players were tiring during the last 15 minutes of a game. When fatigue sets in, muscles tire, work rate slows, and in some respects there is a loss of interest and focus—a situation almost guaranteed to hand victory to the other side. Breaking through the barrier of fatigue can be as much about psychological resilience as physiological endurance, and breath-holding exercises offer a useful technique for improving both.

In order to replicate the conditions of a game, the team's training usually lasted the length of a full match. Their training session consisted of a warm-up and a 10-minute run followed by match practice and tactics. The last 15 minutes included drills and interval training such as running back and forth between cones set at different distances. To incorporate the Oxygen Advantage program as seamlessly as possible into this type of session, I made only modest changes to the existing routine so that the players could adapt to new breathing techniques while maintaining their current form. The result not only increased the effectiveness of the training, but also the players' endurance and performance during competition.

During the 10-minute run at the beginning of training, I instructed the players to switch from their usual habit of breathing through their mouth to running at a comfortable pace with their mouth closed. Every minute or so, each player exhaled and held their breath until they felt a medium to strong air shortage. No changes were made to the match practice portion of the training session, as the introduction of nasal breathing adds an extra load to the body that can initially slow down an athlete, and possibly lead to deconditioning of leg strength. The better approach, therefore, was to incorporate nasal breathing into the 10-minute run and the final 15 minutes of interval training only.

Since the team had a tendency to experience fatigue during the remaining 15 minutes of a game, introducing nasal breathing to the final interval training session added an increased challenge. Running flat out from cone to cone with the mouth closed is no mean feat, and although a couple of the players experienced mild headaches during the first session, the team adapted to it easily. After a few more practice sessions, the players had adapted well to the demands of nasal breathing, so I decided to challenge them further by introducing breath-hold exercises. These exercises (which can be found in the next chapter) subjected the players to an even greater feeling of breathlessness and worked to further delay the onset of fatigue.

Bicarbonate of Soda—More Than Just a Cooking Ingredient!

In a similar way that breath holding delays the onset of fatigue during sports, countless studies have shown that taking the alkaline agent *bicarbonate of soda reduces acidity in the blood to improve endurance.* Who would have thought that a cooking ingredient found in almost every kitchen cupboard in the Western world could also improve sports performance? Not only that, but it is a very helpful tool to reduce your breathing volume and increase your BOLT score.

Bicarbonate of soda is a salt that is found dissolved in many natural mineral springs and is usually sold as baking soda, bread soda, or cooking soda. This ingredient has a wide variety of uses ranging from baking to brushing your teeth to cleaning your fridge.

Taken internally, bicarbonate of soda helps to maintain pH of the blood, and it's also the active ingredient in a number of over-the-counter antacid medications. Dr. Joseph Mercola, a leading authority on natural health, suggests taking bicarbonate of soda for the relief of a number of ailments, including ulcer pain, insect bites, and gum disease.

The therapeutic potential of bicarbonate of soda may soon become more widely known, since Dr. Mark Pagel from the University of Arizona Cancer Center was recently awarded a grant of $2 million from the National Institutes of Health to study the effectiveness of bicarbonate of soda therapy in treating breast cancer.

Over the years many studies have demonstrated the benefits of bicarbonate of soda as a method to help improve sports performance. During high-intensity training, the availability of oxygen for working muscles decreases, which causes an accumulation of acid, leading to muscle fatigue. By ingesting bicarbonate of soda, you can help to maintain normal blood pH by decreasing lactic acid buildup during anaerobic exercise. This alkaline soda neutralizes the acid that accumulates during high-intensity training, resulting in greater endurance and power output.

Bicarbonate of soda can also have positive effects on maximum breath-hold time. As noted throughout this book, improved breath-hold time has positive implications for breathlessness during exercise and your ability to improve your VO_2 max. The ingestion of bicarbonate of soda prior to the practice of breath-hold exercises has been shown to increase maximum breath-hold time by up to 8.6 percent.

For swimmers, the addition of bicarbonate of soda has resulted in an improvement of several seconds during test trials, as well as having significant effects on resting blood pH. Researchers who have investigated the effects of bicarbonate of soda on swimming performance have concluded that the ingestion of bicarbonate of soda can act as an effective buffer during high-intensity interval swimming and could be used to increase training intensity and overall swimming performance. These benefits have even been applied to boxers, leading to improved punch efficacy!

What is consistent throughout all these studies is that the practice of ingesting bicarbonate of soda before exercise successfully neutralizes acid buildup in the blood. In terms of fitness and performance, this means:

- Improved endurance
- Increased maximum breath-hold time
- Reduced breathlessness
- Higher average power output

All in all, quite an impressive array of benefits from a household agent that has no known side effects when taken in small doses!

How to Take Bicarbonate of Soda

I find the following recipe beneficial for improving breathing habits and increasing breath-hold time, and I use it quite often. Try it and take note of its effects on your exercise performance.

You can take bicarbonate of soda an hour or so before training. When you are used to taking it before training, you may also wish to do so before a competition. But like anything else, there is no point in overdoing it. As a precautionary measure, please talk to your doctor before using this approach.

½ teaspoon bicarbonate of soda (also known as baking soda or bread soda)

2 tablespoons apple cider vinegar

1. Put the bicarbonate of soda in a glass.
2. Add the apple cider vinegar and stir for about 1 minute, or until the soda is thoroughly dissolved.
3. Drink the mixture. It will taste a little acidic.

It's as simple as that. Alternatively, you can try drinking ordinary soda water from the grocery store. While traditionally used as a mixer for alcoholic beverages, the carbonation of the water can provide an added effect.

If you drink soda water, please also make sure that you drink your required intake of ordinary still water to ensure adequate hydration. The color of your urine will allow you to determine when you are adequately hydrated; drink enough plain water to ensure it is not too dark, but don't drink so much that your urine is completely clear throughout the day. Drinking too much water is probably just as bad for you as drinking too little. It is about getting the balance right! Until recently, water intoxication or hyponatremia was a little-known and even less understood medical condition. Most people understand that it is sensible to stay hydrated during and after exercise, but when this advice is taken to excess and athletes overdo it, dangerous side effects can result. Marathon runners are particularly susceptible to drinking too much during training and competition—whether water or sports drinks—and this excessive hydration can cause the brain to swell as sodium levels decrease to critically low levels. In a 2002 study of Boston Marathon runners, 13 percent of the runners sampled showed low sodium levels, putting them at risk of serious or even fatal illness. In that very same marathon, twenty-eight-year-old Cynthia Lucero collapsed and died. The state medical examiner's office concluded that the cause of death was a series of medical events brought on by drinking too much fluid during the event. Commenting on the tragedy, Dr. Arthur Siegel of McLean Hospital advised for athletes to weigh themselves prior to a race and write their weight on their race bib. Then, if runners felt unwell during the race, they could be weighed again and treated for dehydration if it was found that their weight was down. If, however, their weight had increased, it would mean that they were overhydrated and that they should retire from the race and stop drinking.

Breath Holding to Prepare for Ascent to High Altitude

Every year, millions of sea-level residents make the journey to high altitudes for recreational skiing and climbing, or for religious, spiritual, or other purposes. Adventurers, climbers, walkers, and sports enthusiasts have ventured to altitudes of over 1,500 meters for the challenge and thrill of the mountains.

British adventurer Bear Grylls reached the summit of Mount Everest in 1998 at the age of twenty-three. In his book *Facing Up,* Grylls describes how he trained for his ascent of Mount Everest by "swimming countless lengths of the local pool—one underwater, then one on the surface, for hours at a time. This boosts one's ability to work without oxygen, making the body more efficient."

There is no doubt that Grylls's training regimen helped his body to acclimatize to the reduced partial pressure of oxygen he would experience during his climb to the top of Mount Everest. Similar to swimming repetitive underwater lengths of a pool, the exercises of the Oxygen Advantage program that simulate high-altitude training can be very helpful to prepare for an ascent to high altitude. More important, as these breath-hold exercises are performed on land, they involve no risk of drowning!

Acclimatization refers to the adaptive changes that the body makes in order to cope with reduced oxygen levels. Most people can ascend to 2,500 meters without difficulty, since oxygen availability is still sufficient at that altitude. However, at higher elevations the oxygen saturation in the blood significantly decreases, making physical activity difficult to sustain.

As you ascend above 2,500 meters, your breathing will become heavier to compensate for the reduced availability of oxygen. Although heavier breathing brings greater quantities of oxygen to the lungs, it also increases the loss of carbon dioxide. As discussed earlier, the loss of carbon dioxide causes blood vessels to narrow and red blood cells to cling on to the oxygen they carry, resulting in reduced oxygenation of tissues and organs. Ironically, as the body breathes more intensely in an effort to take in more oxygen, less is delivered. In a high-altitude environment, oxygenation is more important than ever if altitude sickness is to be avoided.

Almost half of those who attempt to trek or climb to an altitude above 4,000 meters will develop one or two symptoms of mountain sickness after a rapid ascent of more than 400 meters per day. Symptoms vary depending on the individual's physical condition and health and the speed of the climb. Generally, symptoms are mild to moderate and may include:

- Headaches
- Fatigue
- Insomnia
- Loss of appetite
- Nausea or vomiting
- Rapid pulse
- Light-headedness
- Shortness of breath during exertion

A faster climb tends to increase the severity of these symptoms, and may bring about additional symptoms, including:

- Tightening of the chest
- Confusion
- Coughing or coughing up blood
- A bluish discoloration of the skin
- Shortness of breath during rest
- An inability to walk in a straight line

Increasing the oxygen-carrying capacity of the blood is the most important factor when adjusting to an increase in altitude, and breath-hold training exercises are an ideal way to prepare in the weeks before ascent. Spending two to three months performing 5 to 10 maximum breath holds each day will condition the body to accept this intense feeling of breathlessness as a familiar occurrence, potentially resulting in a reduced response to this experience at higher altitude.

Finally, any individual who is serious about his climb should have a basic understanding of how his breathing influences delivery of oxygen to tissues and organs. I can only imagine the number of climbers who intentionally make their breathing more intense to try to counteract the feeling of breathlessness as they climb above 2,500 meters. By now you will be acutely aware that this is exactly the wrong thing to do, and will most likely result in more severe symptoms of altitude sickness. The right thing to do would be to start off with a high BOLT score, to breathe through the nose at all times, and to alter your pace to reduce the feeling of breathlessness.

At least one study shows that breath-hold time is a very useful predictor of mountain sickness and that the lower the breath-hold time, the greater the likelihood of developing symptoms of altitude sickness. In fact, those with a high breath-hold time and a high concentration of hemoglobin in the blood will be better able to tolerate desaturation of oxygen.

Though the ideal BOLT score for each individual will vary, it is reasonable to suggest that a BOLT score of 40 seconds would certainly afford a greater protection from conditions of high altitude than a BOLT of 20 seconds or below.

Prevent Dehydration with Nasal Breathing

The air in mountainous regions and at higher altitudes is cooler and drier than air at sea level. As you ascend to higher altitudes, the increased sensation of breathlessness is likely to induce the switch to mouth breathing. However, since one of the functions of the nose is to moisten and warm incoming air, breathing through the mouth can lead to dehydration, as considerable moisture is expended.

Another factor is that during exhalation, mouth breathing is completely ineffective in retaining moisture. To verify this, gently breathe out through your mouth onto a glass and check the moisture left behind. Now do the same thing, except exhale through your nose. You will find that the moisture left on the glass following nasal exhalation is far less than the moisture left from exhaling through the mouth.

This loss of fluid can contribute to moderate dehydration, resulting in dryness of the lips, mouth, and throat. Other symptoms arising from dehydration include headache, fatigue, and dizziness, which at high altitude could easily be confused with mountain sickness. Heavy breathers and mouth breathers will certainly experience a far greater loss of moisture than those who have a normal breathing volume and who breathe through the nose. Remember that there are no convenience stores at high altitude, so the more moisture you can hold on to, the less you need to carry!

Finally, inhaling cold, dry air through the mouth can cause the airways to narrow. As the airways constrict, the feeling is similar to breathing through a narrow straw, and the result is often to breathe harder and faster to compensate for the restricted airflow. This is a common occurrence experienced by individuals with asthma and can cause even greater dehydration and cooling of the airways, which may lead to even greater respiratory problems.

In the next chapter we will learn exercises that simulate high-altitude training in order to increase the oxygen-carrying capacity of your blood for better sports performance, or to prepare for an ascent to high altitude.

CHAPTER 7

Bring the Mountain to You

World-renowned Brazilian track coach Valério Luiz de Oliveira used breath-hold training techniques with Olympic athletes Joaquim Cruz and Mary Decker, who between them set six world records in 800-meter to one-mile distance running events in the 1970s and 1980s.

De Oliveira's goal was to enable athletes to maintain form at the end of an anaerobic race of 400–800m. Another aspect was to improve psychological preparedness by allowing the athletes to maintain composure during an oxygen-deprived state. One final factor was to train the runners not to pay attention to their breathing, but instead to focus on tactical maneuvers and running form. De Oliveira's techniques were a case of getting results first and figuring out the science later, but his theories certainly proved to be right.

The method de Oliveira used with his athletes is as follows:

- Athletes run 200 meters on a straight course at near race pace, holding the breath on an inhalation for the last 15 meters.
- After a partial rest of 30 seconds, this breath hold is repeated 3 more times.

- Athletes then spend 3 minutes recovering before repeating.
- In total, the athletes perform 3 sets of 4 breath holds.

According to de Oliveira, "Everybody's capable of holding their breath for a very long time. But you've got to do three of these sets. By the final set, you're going to become very, very tired. It's hard to hold your breath at that point. But if you use my drill, you will see results."

De Oliveira uses another exercise in which his 400m and 800m runners hold their breath for the last 30 meters, simulating the end of a race when they will be most fatigued. Maintaining form during the last 30 meters of a race like this is crucial. According to de Oliveira, "The most important thing you can do in the race no matter how exhausted you get is to maintain your form."

Joaquim Cruz, coached in this way by de Oliveira, won gold in the 800m competition during the 1984 Olympics in Los Angeles and silver at the 1988 Olympics in Seoul. Further accomplishments included bronze in the 800 meters at the 1983 World Championships and gold in the 1,500 meters during the 1987 and 1995 Pan American Games. By the end of 1984, he was the National Collegiate Athletic Association champion and Olympic champion; undefeated in all seven of his 800m finals; had run the second-, fourth-, fifth-, and sixth-fastest 800m times in history; and easily ranked as number one in the world for the 800 meters in 1984.

The legendary Czech athlete Emil Zátopek, described by the *New York Times* as perhaps one of the greatest distance runners ever, also incorporated breath holding into his regular training. Zátopek was a man of small stature, standing 5 feet 8 inches tall and racing at 139 pounds, but he found an edge over his competitors by developing his own innovative training techniques, which included interval training and breath holding. While walking to and from work each day, he passed a lane lined with poplar trees. On the first day, he held his breath until he reached the fourth poplar. On the second day he held his breath until he reached the fifth poplar, increasing the distance of his breath hold by one tree each day until he could hold his breath for the entire line of trees. On one occasion, Emil held his breath until he passed out. While it is not necessary to practice to this extreme, it is fascinating to note that one of history's greatest runners was performing breath holding long before it formed part of training for some present-day athletes.

The difficulty in presenting accounts of active athletes is that they strive to keep their training regimens a secret. It doesn't make much sense to disclose your innovative techniques, especially when they give you an edge over the competition. To date, I have worked with a number of Olympic and professional athletes who have incorporated the breathing exercises in this book into their training. Given the tiny (but highly important) difference in margin of performance between one elite athlete and another, I am very much aware of the importance of keeping training information close to one's chest.

However, sometimes hints and snatches of information can be gleaned through the press, suggesting that breath holding is becoming more widely practiced in the athletic community. For example, a recent article on the athletic website Eightlane.org reported that Galen Rupp—the current American record holder for the 10,000 meters and indoor 3,000 meters, and silver medal winner at the 2012 London Olympics—had recently collapsed during training. Rupp's headphones had fallen off and "he was unable to hear his coach reminding him to breathe." Reading between the lines, it seems as if Rupp may well have been practicing breath holding during his training sessions, with his coach setting his limits. Please note that it is neither necessary nor safe to hold your breath in order to create such an extreme air hunger. To get the most benefit from the breath-hold exercises in this book, only hold your breath until you experience a medium to strong urge for air. Pay attention to the intensity of the air hunger you create as you perform each exercise, and release your nose when you feel challenged. You should be able to recover your breathing within 2 to 3 breaths following a breath hold.

To the uninitiated, purposely holding the breath may seem strange. Oxygen is necessary for life, so why subject the body to such a limitation? Just as physical training is a very natural activity for mankind, so too is breath holding. As a child, you might have held your breath any time you swam to pick up a coin or other object from the bottom of a swimming pool. At other times, you may have held competitions with siblings or friends to see who could hold their breath for the longest, with passing the 1-minute mark being an acceptable accomplishment.

For the past thirteen years, thousands of children have attended my courses to help address coughing, wheezing, and symptoms of breathlessness and asthma. Children as young as four are able to practice a number of different breath-hold exercises, each with a specific purpose. For example, there is a breath-hold exercise to help unblock the nose, another to help stop a wheeze or a cough, while another is designed to improve breathing volume by holding the breath for as long as possible.

While adults may at first be wary of holding their breath, children often take to it like ducks to water. I usually work with five or six children to a group, ranging in age from four to fifteen years. Beginners are gently introduced to the exercises by walking a distance of 10 paces while they hold their breath. After 3 or 4 repetitions, the number of paces is increased in increments of 5 until the child understands the exercise and experiences a moderate need for air. Most children master the exercise in no time at all, and are soon in friendly competition with their peers to hold their breath for as many steps as possible.

I usually expect children to hold their breath for 30 paces during the first session, increasing by 10 extra paces each week. Some children will make quicker progress and can learn to hold their breath for up to 80 paces in as little as two or three weeks without losing breath control or experiencing any stress. Even professional athletes would be impressed by this feat. More important, at 80 paces, in my experience, the child's blocked nose, cough, wheeze, or exercise-induced asthma will have completely disappeared.

The beauty of breath holding is that while the air shortage can be relatively extreme, it is entirely under our own control and for a short duration of time only.

For premenstrual women, vegetarians, or those with a history of anemia, it may be necessary to take an iron supplement to support the production of normal red blood cells. If, despite sustained practice of these exercises, your BOLT score fails to increase, it may be useful to visit your doctor to have a complete blood count. If your hemoglobin is low, speak with your doctor about iron supplementation. In some individuals, I have witnessed iron supplementation making a remarkable difference to BOLT scores within just a few short weeks.

The following Oxygen Advantage breath-hold training exercises provide simple ways to simulate the beneficial effects of high-altitude training and high-intensity training while allowing you to exercise in your usual way. Each exercise provides both a hypoxic (lack of oxygen) and a hypercapnic (high carbon dioxide) response. Combining these two effects produces important changes in the body, such as:

- Lowering sensitivity to carbon dioxide
- Increasing endurance
- Reducing discomfort and fatigue from lactic acid buildup
- Increasing the oxygen-carrying capacity of the blood
- Improving breathing economy
- Improving VO_2 max

By incorporating these easy techniques into your routine, you will find that your breath-hold ability increases quickly, and you will begin to see results within your training and your competitive performance.

Using a Pulse Oximeter

To get the best results from breath-hold exercises, it can be helpful to use a handheld device called a pulse oximeter that measures how loaded the blood is with oxygen. Convenient and noninvasive, using a pulse oximeter simply involves placing a probe on your fingertip to measure the oxygen saturation (SpO_2) of your blood. A pulse oximeter can be purchased rather cheaply, but my advice is to purchase a better-known brand such as NONIN, as they tend to be more responsive. One of the main benefits of using a pulse oximeter is that it can be very motivating to witness the drop in oxygen saturation as you practice breath holds, reinforcing the success of the exercise. In addition, the device can help to ensure that you don't overdo the exercises by lowering your SpO_2 below 80 percent.

Normal oxygen saturation at sea level varies between 95 and 99 percent (as we have seen), while the benefits from breath holding occur when the SpO_2 level is dropped to below 94 percent. In the beginning, you may not notice much of a decrease to your oxygen saturation while performing breath holds. However, with practice and the ability to tolerate a stronger air shortage, the drop to oxygen saturation will become evident in as little as a few days.

The effects of breath holding depend on two factors—oxygen saturation during training and the length of the exposure to reduced oxygen—but slow and steady is the way to go, just as it is whenever you start a new exercise program. To get the most out of breath-hold exercises, it helps to start gently, holding the breath until you feel a medium air hunger during the first two or three breath holds before gradually increasing the duration and intensity. This way, you will feel in control and be able to practice more effectively. As your BOLT score increases, you will find it easier to manage the sensation of air hunger, and your blood oxygen saturation will start to drop below 94 percent.

Simulate High-Altitude Training While Walking

We begin by introducing a simple walking exercise that enables you to achieve similar benefits to those experienced during intense physical training in as little as 10 to 15 minutes. The beauty of this exercise is that it can be performed anywhere and at any time, even if an injury is preventing you from engaging in normal training. Similar to any intense physical exercise, it is important to practice at least two hours after eating. Just as it is not advisable to go for a jog directly after eating, it is also best to practice breathing exercises on an empty stomach. Not only would it be uncomfortable to perform breath holds too soon after a meal, but the benefit of the exercise would be much reduced as the process of digestion increases breathing.

During this exercise you will be practicing breath holds as you walk. For the first 2 to 3 breath holds, in order to gently acclimatize your body to lower levels of oxygen, it's important to hold your breath only until you feel a medium hunger for air. For the remaining breath holds, challenge yourself by aiming to achieve a relatively strong need for air. Due to a delay in the pulse transit time, it is common for the decrease in oxygen saturation to take place not during the breath hold, but soon after it. Therefore, to get the most from the exercise, minimize breathing for about 15 seconds following the breath hold by taking short breaths in and out through the nose. If you have a pulse oximeter, you might enjoy observing the decrease to your oxygen saturation as you do this—effectively simulating high-altitude training and bringing the mountain to you.

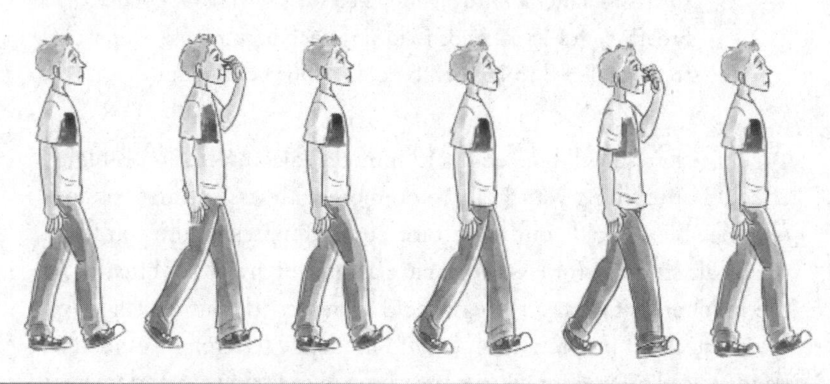

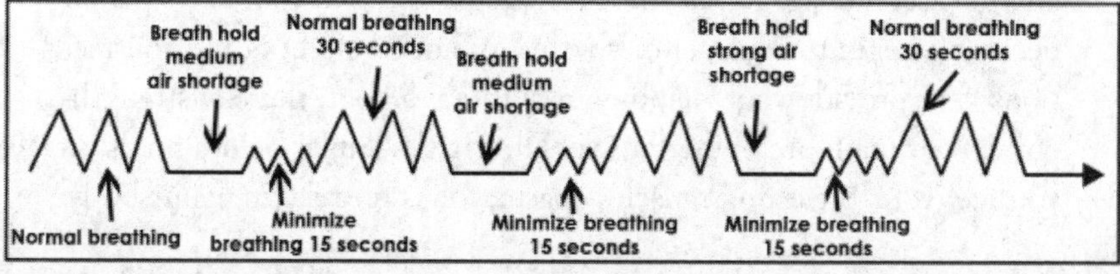

- **Walk and hold:** After a minute of continuous walking, gently exhale and pinch your nose to hold your breath. If you feel uncomfortable pinching your nose while walking in public, you can simply hold your breath without holding your nose. Continue to walk while holding your breath until you feel a medium to strong air shortage. Release your nose, inhale through it, and minimize your breathing by taking very short breaths for about 15 seconds. Then allow your breathing to return to normal.
- **Continue walking for 30 seconds and repeat:** Continue walking for around 30 seconds while breathing through your nose, then gently exhale and pinch your nose with your fingers. Walk while holding the breath until you feel a medium to strong hunger for air. Release your nose and minimize your breathing by taking short breaths in and out through your nose for about 15 seconds. Then allow your breathing to revert to normal.
- **Repeat breath holds 8 to 10 times:** While continuing to walk, perform a breath hold every minute or so in order to create a medium to strong need for air. Minimize your breathing for 15 seconds following each breath hold. Repeat for a total of 8 to 10 breath holds during your walk.

This exercise will take about 12 minutes to complete and is highly effective at teaching your body to do more with less. At first you may only be able to hold your breath for 20 or 30 paces before you feel a strong air shortage (or less if you have asthma or are out of breath). As the number of paces per breath hold increases, the air shortage you experience will progress from easy to moderate to strong. As you feel an increased hunger for air, the breathing muscles in your abdomen or neck will begin to contract or spasm. An added effect of the contractions is to provide your diaphragm with a workout, thereby strengthening your main breathing muscle. During the longer breath holds, as you feel your breathing muscles spasm, focus on relaxing your body.

Allow your muscles to go soft as you hold your breath. Relaxing the body in this way allows a longer breath hold with less stress.

With repetition, as the weeks go by, you will find yourself being able to hold your breath for 80 to 100 paces. Your ability to hold your breath will increase with practice and without stressing your body. Do not overdo it. Ideally, your breathing should recover easily and become calm within 3 or 4 breaths. While this exercise is a challenge, it should not be stressful.

If you notice any side effects, such as an elevated or stronger than normal pulse for a prolonged period after completing breath holding, then it is best to refrain from performing the stronger breath holds. Instead, concentrate on breathing lightly both during rest and physical exercise to bring benefit to your health and sports.

Breath holding can also be incorporated into a jog, run, or bike ride. While you may not be able to hold your breath for as many paces during a jog as you can during a walk, the quality of the exercise will be better because of the greater accumulation of carbon dioxide in the blood.

Breath holding during training adds an extra load that would only otherwise be experienced during maximum intensity exercise.

Here is a breath-hold exercise to try while jogging or running:

- **Run and hold:** Ten to fifteen minutes into your run, when your body has warmed up and is sweating, gently exhale and hold your breath until you experience a medium to strong air shortage. The length of the breath hold may range from 10 to 40 paces and will depend on your running speed and BOLT score.
- **Break for 1 minute and repeat:** Following the breath hold, continue to jog or run with nasal breathing for about 1 minute, until your breathing has partially recovered.
- **Repeat breath holds 8 to 10 times:** Repeat the breath hold 8 to 10 times during your run, followed each time by a minute of nasal breathing. The breath holds should be challenging but should also allow breathing to recover to normal within a couple of breaths.

If you find this exercise in any way stressful, or have difficulty recovering your breathing after a breath hold, then refrain from doing the exercise until your BOLT score has increased to at least 20 seconds.

Breath Holding During Cycling

A similar practice can be employed as you ride your bicycle:

- After your body has warmed up, exhale and hold your breath for 5 to 15 pedal rotations.
- Resume nasal breathing while continuing to cycle for about 1 minute.
- Repeat this exercise 8 to 10 times throughout your ride.

Breath Holding During Swimming

Swimming is the only sport in which breathing volume is controlled, as the face is immersed in the water and the weight of the body on the water restricts breathing even further. Breathing through the mouth is probably the best option during swimming, as nasal breathing may result in water inhalation.

To incorporate reduced breathing during swimming, you will need to increase the number of strokes between breaths. You can do this in gradual increments, increasing the number of strokes between breaths from 3 to 5 to 7 over a series of lengths. This exercise was employed by former Olympic swimmer and triathlete Sheila Taormina, who achieved the fastest 1.5 km swim during the Sydney Games in 2000. In my correspondence with Taormina, she explained how reduced breathing was used to create a training effect so that the swimmer would be challenged to do more with less. However, unlike some sports, such as underwater hockey, the breath hold is never pushed to an absolute maximum, in order to ensure the safety of the swimmer.

In addition to the hematological benefits from breath-hold training, it has been found that breath holds also improve swimming coordination. After breath-hold training, swimmers have shown increases in VO_2 peak as well as an increase in the distance traveled with each swimming stroke. During the front crawl it is necessary to turn the head sideways every few strokes in order to take a breath. However, each time a swimmer takes a breath, hydrodynamic drag takes place, which ultimately wastes energy and reduces performance. One of the benefits of having a high BOLT score is that breathing becomes more efficient, meaning that less air is required during a swim. Reducing the number of breaths minimizes drag, helping to conserve energy for improved performance.

The same training has also been applied to underwater hockey players, who perform their sport below the surface of the water in a swimming pool. The object of the game is to pass a weighted puck along the bottom of the pool using an underwater hockey stick, scoring points in the opposing team's goal. As play takes place underwater, it is advantageous, if not essential, that players are able to hold their breath for long periods of time. Part of the training of underwater hockey players includes prolonged and repeated breath holds and controlled breathing, which lead to a higher tolerance for carbon dioxide and an increase in breath-hold time.

Researchers investigating the effects of short repeated breath holds on underwater hockey players found that these exercises reduced breathlessness and produced a higher concentration of carbon dioxide in the blood. In addition, lactate values were found to be lower in underwater hockey players compared to untrained individuals, meaning that the pain from lactic acid buildup was reduced. These athletes clearly have a high tolerance for carbon dioxide, most likely explained by their experience with prolonged breath holding for their sport. As we have already seen, reduced sensitivity to carbon dioxide translates into reduced breathlessness during exercise as less breathing is required to eliminate excess carbon dioxide. This allows greater physical exertion with lighter breathing—enabling your body to do more with less.

Advanced Simulation of High-Altitude Training

Normal oxygen saturation at sea level varies between 95 and 99 percent. To receive any benefit from hypoxic (reduced oxygen) training, oxygen saturation levels must drop below 94 percent (and ideally to below 90 percent). The effect of this method depends on two factors: oxygen saturation during training, and the length of the exposure to reduced oxygen.

Lowering oxygen saturation below 90 percent for a duration of 1 to 2 minutes can significantly increase the production of EPO, and this can easily be achieved by using this exercise.

Before you consider doing this exercise, *please get the all clear from your physician.* This exercise is only suitable for those who have good physical fitness, perfect health, a BOLT score of longer than 30 seconds, and are accustomed to performing intense physical exercise. In other words, if you wish to try this exercise, you must be familiar with experiencing strong air shortages. Please refrain from doing this exercise if any of the following statements apply to you:

- You are in any doubt about your capabilities to perform intense physical exercise.
- You are unwell.
- Your BOLT score is shorter than 30 seconds.
- You are not currently in a regular physical training program.

This advanced exercise aims to readjust the composition of your blood and alter levels of oxygen and carbon dioxide. After months of experimenting, I have developed this exercise to lower the oxygen saturation of arterial blood and to maintain this decrease over a number of seconds. I have practiced this exercise hundreds of times and include the following guidelines to help you to perform it correctly and to be aware of potential side effects:

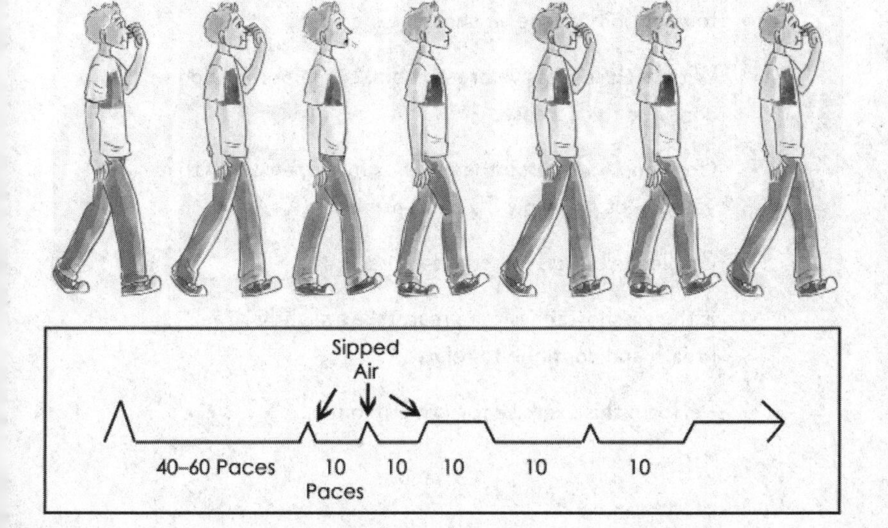

- To regulate the decrease in oxygen saturation to below 94 percent and to ensure that it does not go below 80 percent, it is important to use a higher-quality pulse oximeter during this exercise.
- This exercise should be practiced on a relatively empty stomach, at least three hours after eating.
- The first breath hold is between 40 and 60 paces, or until you feel a medium to strong need for air.
- After the first breath hold, subsequent holds are performed every 5 to 10 paces.
- Following each breath hold, either exhale through your nose or take a sip of air in through your nose before the next breath hold.
- A "sip of air" means taking a tiny breath in, the purpose of which is to relieve tension rather than take in air. It is about 10 percent of a normal breath.
- Contractions of the diaphragm will strengthen as the air shortage progresses. Try to bring a feeling of relaxation to your body as the air shortage increases.
- With each successive breath hold, oxygen saturation will continue to decrease.
- Continue to observe the pulse oximeter, ensuring that you do not go below 80 percent SpO_2.
- Challenge but do not stress yourself.
- If the air shortage is too great, take a slightly larger breath and continue to relax.
- Perform this exercise for 1 to 2 minutes.

The objective of the exercise is to generate a reasonably strong air shortage in order to lower oxygen saturation and to maintain it at a lower level for a period of between 30 seconds and 2 minutes.

Please note that it is not advisable or even necessary to lower your oxygen saturation below 80 percent. Maintaining an oxygen saturation of less than 91 percent for approximately 24 seconds can result in an increase of EPO of up to 24 percent, while maintaining this saturation for 136 seconds can result in an increase of EPO of up to 36 percent.

Putting the Oxygen Advantage Program to the Test

France is well known for its global cycling events. It is not only the home of the Tour de France but also to such legendary mountains as Mont Ventoux and Col le da Madone, which lure cyclists from around the world to conquer their slopes. The thriving French amateur road racing culture is steeped in folklore, setting such intense challenges that some athletes are unable to compete more than a few seasons due to injury, fatigue, or burnout.

Nick Marshall is an Australian cyclist who started racing in the Paris region. A father of two and a businessman, Nick had trouble juggling the demands of work and family with what he felt were “old-school training methods.” Seeking a better way to train, Nick started using advanced yoga breathing and then the Oxygen Advantage principles to develop more power while reducing his overall training load. Initially, Nick had a BOLT score of just 25 seconds (common even among elite athletes), but through the nasal breathing, Breathing Light to Breathe Right, and simulating high-altitude training, his BOLT score now regularly reaches 60 seconds.

The primary Oxygen Advantage exercise Nick used was a daily 30-minute routine combining the following:

1. Breathe Light to Breathe Right for 15 minutes.
2. Simulate High-Altitude Training while walking with breath holds of 60 to 80 paces.
3. Rest for 3 to 4 minutes.
4. Do 1 set of Advanced Simulation of High-Altitude Training to reduce arterial blood oxygen saturation to around 81 to 84 percent.

The application of Oxygen Advantage techniques has resulted in a drop in lap time along with a reduction in weight and the improvement of Nick's overall health. On his bike, Nick has seen an improvement in VO_2 max and lactate buffering during hard efforts, and a reduced heart rate at rest. (Lactate buffering refers to the body's ability to offset or neutralize the effects of lactic acid during intense exercise.) A big benefit for Nick is the ability to reduce the training hours spent on his bike while still steadily improving his fitness—a clear sign that he has become more efficient as an athlete.

Oxygen Advantage Training: Short- and Long-Term Benefits

Oxygen Advantage training involves temporarily subjecting the body to reduced oxygen saturation. This is usually achieved by living or training at high altitudes, but breath-hold exercises can easily bring about the same result. Performing just 5 maximum breath holds can significantly increase the concentration of oxygen-carrying red blood cells in the blood, but often levels will return to normal within 10 minutes following the final breath hold. Does this mean that breath-hold training is only beneficial directly before a competition? The answer is no. Several research studies have shown that permanent improvements to oxygen-carrying capacity can be achieved by regularly exposing the body to reduced concentrations of oxygen. If you incorporate the Oxygen Advantage program into your normal training and practice nasal breathing during rest and day-to-day activity, you will begin to see real physiological changes that will enable you to raise your game and increase your endurance over both the short and long term.

The benefits of genuine and simulated high-altitude training have been studied in many different ways, and the results repeatedly show that long-term exposure to reduced oxygen concentration brings about advantageous changes for individuals looking to improve their physical performance.

Breath-hold divers have been found to show a 5 percent higher resting hemoglobin mass than untrained divers, suggesting that long-term breath holding has a tangible effect on performance. In addition, experienced breath-hold divers demonstrate a stronger spleen contraction in response to breath-hold exercises, leading to a greater release of red blood cells into the blood supply, improving their oxygen delivery.

In chapter 1, I mentioned Don Gordon, who has reached new heights in his cycling performance by applying the Oxygen Advantage program. In a recent e-mail, he informed me that his hematocrit was 52 percent (up from 47 percent)—high enough to provide increased aerobic performance while still remaining within the upper normal range.

As I have noted before, it is not possible or practical for everyone wishing to benefit from the effects of hypoxic training to live at high altitude, but fortunately beneficial results can be obtained with less drastic alterations to your lifestyle and training routine.

The adage "use it or lose it" can be applied to all forms of training; form, fitness, and endurance can only be maintained through constant repetition and practice. The same is true for your breathing. First you must learn how to breathe efficiently and correctly day and night, during rest and activity. Only then should you apply Oxygen Advantage techniques to your training routine and sporting competition. By practicing these exercises regularly, you will benefit from all the advantages of high-altitude training, increase your VO_2 max, and move beyond your previous limits.

To fully utilize the power of Oxygen Advantage reduced breathing exercises, it is important to bring a feeling of relaxation to your body as you exercise, and encourage your breathing volume to reduce. Exercise at a pace in which your breathing is regular and controlled, and you experience a feeling of air shortage. To generate greater intensity and bring about positive physiological changes from hypoxic training, incorporate breath holds into your physical training. Your BOLT score will provide continuous feedback on your breathing volume during rest and physical exercise. If your BOLT score decreases, it means your breathing is heavier than your metabolic requirements. This will have negative implications for both performance and health. Return to the earlier exercises and concentrate on retraining your breathing habits until you are able to improve your BOLT score to more than 30 seconds.

When you first begin these exercises, your breath-hold time and BOLT score may be short, but with regular training and a commitment to the program, both these measurements can be increased over a short period of time.

A number of research papers have come to conclusions that support my findings, studying athletes and nonathletes alike to see the effects of breath-hold exercises on an individual's tolerance of carbon dioxide. These studies have found that breath-hold time can be increased as a result of short-term and long-term practice. For example, a study measuring the breath-hold time of volunteers practicing breath-hold exercises with their faces immersed in water found that the length of breath holds increased by up to an impressive 43 percent over a series of exercises. Another study found that breath-hold divers with 7 to 10 years of experience were able to hold their breath for up to 440 seconds, compared to up to 145 seconds achieved by inexperienced individuals. Similarly, the duration of breath-hold time for triathletes was found to be significantly lengthened following a three-month breath-hold program.

Earlier on, we investigated the effects of a reduced sensitivity (ventilatory response) to carbon dioxide, finding that it resulted in better sports performance, reduced breathlessness, and improved VO_2 max. Your BOLT score allows you to measure this sensitivity and provides a useful method for tracking your progress, providing you with both a goal to strive for (attaining a BOLT score of 40 seconds) and positive reinforcement that you are making real changes to your body's capabilities as you watch your score increase. All the exercises in this book aim to increase your BOLT score and improve your body's tolerance for CO_2, and you will be able to notice improvements to your well-being and performance right from the very first time you practice.

CHAPTER 8

Finding the Zone

Considered the greatest moment in twentieth-century boxing, the 1974 "Rumble in the Jungle" pitted undefeated world heavyweight champion George Foreman against former champion Muhammad Ali. The event was organized by boxing promoter Don King and sponsored by the King of Zaire with the promise of a major purse to the winner.

No one thought Ali had much of a chance of beating Foreman. After all, Foreman was both younger and larger than Ali and was considered the strongest fighter of his generation—no previous opponent had lasted more than three rounds with him. But Ali had more than just speed and strength—he used psychology and tactics to his advantage. During the early rounds of the fight Ali toyed with Foreman by frequently leaning on the ropes and covering himself up, leading Foreman to throw ineffective body punches and tire himself out. By the seventh round, Ali turned to taunting, goading the weary Foreman with jibes like "They told me you could punch!" and "That all you got, George?"

In the eighth round Ali saw his moment and took it, landing a strong left hook and hard right. Foreman, weakened by fatigue and distraction, stumbled to the canvas, and though he managed to get up at the count of nine, the referee called the bout to an end. Muhammad Ali, a master in psychology, won the title by a knockout.

Very few people could have expected this outcome. Both fighters were equally motivated to win, but while Foreman was at that time the stronger combatant, Ali's constant barrage of taunts played a significant role in undermining Foreman's mental strength, causing his concentration to lapse and allowing him to lose his temper—the opportunity Ali needed to strike. By pulling his opponent out of the zone, Ali created an opening for himself, overcoming all odds. And that's all it takes—just one distraction can dramatically alter the outcome of any event. Often athletes are deprived of success not because of lack of skill, fitness, or stamina, but by their own thoughts.

When looking back on a disappointing performance, most athletes will comment that their head "just wasn't in it." Training the mind to be in the flow is just as vital as training the body. As any athlete knows, one thought is all it takes to divert attention from the task at hand, ruining the shot, penalty, race, or putt. But while in the flow, distracting thoughts do not enter. The shouts of opposing spectators are not heard, a mistake made during the game is not ruminated on, and thoughts of past mistakes or of future goals do not arise. There is no fear of losing. There is no expectation of winning. You are not anxious about actions or reactions from opponents but effortlessly perform to the best of your ability. Nothing else matters. You are present, using the full concentration of your mind in a state of undivided attention.

"Going with the flow" is a concept popularized by Mihaly Csikszentmihalyi, former head of the department of psychology at the University of Chicago. Csikszentmihalyi described "flow" as the experience of "being completely involved in an activity for its own sake. Self-consciousness falls away. Time flies. Every action, movement, and thought follows inevitably from the previous one, like playing jazz. Your whole being is involved, and you're using your skills to the utmost." This mental state is also sometimes described as being in the zone or the present moment.

Flow is a state of concentration that allows for complete immersion in the situation at hand. Being in the flow means that no boundaries exist between you and the activity in which you are involved. The player and the game become one. The ego—which is the fictitious story that we create about ourselves—is left behind. Conscious thinking ceases, and the athlete acts spontaneously. Any sense of self-consciousness is set aside, allowing full concentration and focus to be obtained. While in the flow, instinct and intuition take over and the right action happens automatically, without the need for conscious thought.

When in the flow you do not think about how good you are, or how useless you are, or what the spectators think of you, or what you are going to do tomorrow, or what your hair looks like. The usual repetitive nonsense generated by the active Western mind ceases. Concentration—the ability to focus unhindered by distracting thoughts—is at its highest. In such a state of intense concentration, your complete attention can be devoted to the game.

Being in the flow allows for a still, quiet mind, undistracted by conscious thoughts. It is a state that involves the use of the entire brain rather than just the logical processes of the left brain. Being in the flow is the very antithesis of Western education, whose sole purpose is to develop and nurture the analytical, reasoning, and logical brain.

You will no doubt have experienced the feeling of an activity taking up all your attention and focus to the point where you forget everything else around you. When you are truly engaged in creative endeavors like sports, writing, painting, music, and drama, many hours can pass by unnoticed. The dancer and the dance become one. The painter and the painting become one. The runner and the race become one.

During training practice, an athlete will perform the same action over and over again, making precise adjustments to create the perfect sequence, whether it is marking an opponent during a game, taking a golf swing, knowing when to overtake during a race, or taking a penalty kick. In the same way, a martial arts expert develops controlled and flawless movements through years of constant repetition and refining. Each time a movement is replicated, the brain stores information and develops muscle memory, eventually allowing the task to be performed without conscious effort. In essence, the body knows what to do—the mind simply needs to get out of the way. There is no room for thought in fast-paced activities; thoughts serve only as distractions. An athlete at peak performance does not think. Instead, instinct kicks in, muscle memory ensures spontaneous movement, and 100 percent of energy is directed toward the task at hand. In the zone, the athlete finds that his or her reactions and movements flow without conscious thought. Intuition takes over and the right action occurs naturally.

During the 1988 Monaco Grand Prix, Ayrton Senna managed to stay ahead of his competitors with ease, including a teammate driving a similar car. Recounting the race, Senna explained how he achieved this without conscious effort: Letting his instincts guide him, he felt as if the circuit had become a tunnel, and no matter how fast he drove, there was always room for more.

Enter the Zone at Will

The zone is nothing more than performing in the absence of thought. When the mind is still and thoughts are no longer a distraction, you are able to give an activity your complete and undivided attention. Being able to focus without distracting thoughts defines concentration, a vital attribute for the proper execution of any action and the achievement of any goal. An active mind results in diminished concentration as every little thought interferes with the task at hand. For example, an individual who is reading this book with an active mind is merely looking at the page. His attention is elsewhere on a never-ending train of repetitive and useless thinking. Although his eyes might be following the sentences, his mind is not. When he reaches the bottom of the page, it is unlikely that he will remember much of the content.

Nowadays, as we spend more time communicating via social media, playing computer games, and surfing the Internet, our powers of concentration are diminishing. According to international motivational guru Kevin Kelly, we are now living in an attention-deficit society. The dial has moved from conversation to presentation and from dialogue to monologue. We no longer give each other our undivided attention, and neither do we take the time to observe our own breathing or allow our minds to still.

Ted Selker of the Massachusetts Institute of Technology supports this view, claiming that because the Internet offers so much choice we end up spending our time flitting from one thing to another, shortening our attention spans and forming a habit of poor concentration. Selker suggests that the nature of web browsing can leave us with an attention span of just 9 seconds—the same as that of a goldfish.

Reading a piece in the *New York Times,* I was surprised to learn that the children of the late Apple founder Steve Jobs hadn't been introduced to the iPad. When journalist Nick Bilton asked Jobs if his children loved the gadget, he replied, "They haven't used it. We limit how much technology our kids use at home." This stance has been echoed by a number of technology chief executives, who set very strict criteria on the amount of screen time their children experience—all too aware of the implications of extended periods spent looking at a screen. As the modern world becomes increasingly dependent on screens and devices, gadget addiction has the potential to isolate people from the world around them, subduing social interaction and increasing mind activity.

Not only is an overactive mind less conducive to focus and productivity, it also leads to increased stress, anxiety, and depression—all of which contribute to mental health problems and a reduced quality of life.

The importance of being able to control and still the mind cannot be overestimated. An athlete with a quiet mind will enjoy good powers of concentration and be able to enter the zone at will, but an athlete with an active mind will have a head full of unnecessary thoughts and will find it difficult to enter the zone. If the mind is overactive during daily life, it follows that the mind will be overactive during sports. Only when the mind is relatively still during normal daily life can the athlete enter the zone during competition. A still mind can be attained through having a high BOLT score, using meditation, and developing awareness of the mind—nothing else.

You could, of course, go down to your local bar and drink six or seven large beers. No doubt this approach will quiet the mind. As appealing as it sounds, achieving a state of mind where thoughts are drowned out with alcohol is not conducive to alertness and increased performance. Instead, for thousands of years, human beings have practiced different forms of meditation to quiet and tame the mind. Meditation allows you to pay attention to your thoughts, emotions, and feelings, while reducing repetitive and useless thinking.

Ryan Giggs made his first appearance for the Premiership football club Manchester United during the 1990–91 season. As the most decorated player in English football history, Giggs has won thirteen Premier League medals, four FA cup winner's medals, three League cup winner's medals, and two Champions League winner's medals. At the age of forty, when most of his peers had long since retired, Giggs continued to play professional Premier League football. So what is his secret? According to Giggs, self-awareness has been a major factor in his extended career: "The focus on oneself is hugely important, even if it is only for a daily hour of stretches and meditation," he says.

The renowned golfer Tiger Woods is also well known for using meditation to improve his game. Tiger's father, Earl Woods, was instrumental in developing his son's concentration. Earl Woods explained he would repeatedly attempt to distract his son as he practiced his swing by dropping a golf bag or shouting obscenities. Earl Woods believed that Tiger could be the "first black intuitive golfer ever raised in the United States" and tested his meditative concentration from a very young age. And his prediction came true: Tiger Woods has been the world number one golfer for more consecutive weeks than any other player in history. The ability to play sports intuitively relies on complete immersion in the zone, where the right action occurs effortlessly and the golfer and the game become one. In the film *The Legend of Bagger Vance,* the celebrated coach describes the perfect swing as being in harmony with "all that is, all that was, and all that will be."

To get to this place and access intuitive intelligence, it is necessary to practice quieting the mind. Intuitive intelligence is not learned but experienced. Those who bring about monumental change and success in this world have access to it. For some it is an automatic process. For others, including myself, it has to be developed. A clear example of intuitive intelligence can be found in the late Steve Jobs. In an interview with his biographer Walter Isaacson, he described observing how people living in India relied a lot more on their intuition than the analytical reasoning of the West. Jobs believed that intuitive intelligence was more powerful than the intellectual intelligence revered in the Western world. Jobs was a dreamer, putting logic aside and accessing the power of universal intelligence through a still mind. It is because of this intuition and creativity that products such as the iPhone, iPad, and Mac came into being.

In the past, meditation has evoked a negative image as a hippie exercise practiced by those with nothing better to do. Slowly but surely this image is changing, as scientists have begun to recognize the many benefits of quieting the mind, such as reducing anxiety and improving attention and concentration during stressful challenges.

A 2014 study investigated whether mindfulness meditation–based techniques influenced the resilience of United States Marines. Eight Marine infantry platoons comprising 281 soldiers were randomly divided into two groups. One group received twenty hours of classroom instruction in mindfulness and practiced the techniques for at least a half hour daily for eight weeks. The second group did not receive any mindfulness training. Both groups then participated in training under battlefield conditions. In a report published in the *American Journal of Psychiatry,* the researchers concluded that the marines who practiced mindfulness experienced improved quality of sleep, reduced stress, and a quicker recovery of heart rate and breathing following intensive combat training.

In other studies with U.S. Marines trained in mindfulness-based techniques, brain scans revealed similarities to those of Special Forces soldiers and Olympic athletes, and the area of the brain responsible for controlling fear was shown to have actually shrunk. During warfare, business, sports, or even normal family life, a calm, attentive, and collected mind always results in better decision making. Staying fully focused during a stressful situation is essential if you want to take the correct course of action.

Until recently, it was commonly believed that the brain stopped developing when we reached adulthood. It has only been within the last few years that scientists have discovered that the brain can be changed by practicing mindfulness meditation. This is tremendous news, not only for those in sports but also for anyone suffering from anxiety or depression. The ability to bring about changes in the brain could enable people to take back control of their mental health rather than accepting a lifelong reliance on mind-altering medication.

Neuroscientists from the most prestigious universities around the world, including Harvard and MIT, have conducted much research to investigate the changes that take place in the brains of people who meditate. There is strong evidence that present-moment awareness literally changes the brain, making many regions more powerful and efficient. A team of scientists from the University of British Columbia and Chemnitz University of Technology pooled data from twenty studies investigating this phenomenon. Remarkably, all studies showed that mindfulness meditation resulted in an increased density of gray matter, leading to more effective processing of information. MRI scans revealed that at least eight different regions of the brain increased in efficiency, including the orbitofrontal and hippocampal, which play a role in our ability to hold attention, cultivate positive emotions, and retain emotional stability. People who meditate are happier with themselves, are better able to resist distractions, and learn from past experience—all traits that are essential in today's modern life.

The ancient Greek aphorism "know thyself" is written in the forecourt of the Temple of Apollo at Delphi. While its exact meaning is often the subject of debate among scholars, in the context of harnessing the power of meditation, the words ring with truth. When you are mindful, you become more aware of your inner monologue, allowing you to stop the cycle of compulsive thought processes and step out of the prison of self-doubt. Only when you realize that you are a prisoner of the mind is it possible to escape. And while the walls and bars may not be concrete and steel, being a captive of your thoughts can have a significant effect on your ability to focus and perform.

I like to divide thoughts into two categories: *practical thoughts,* which are useful and serve a purpose, and *distracting thoughts,* which serve no purpose at all. To decide on a course of action and achieve anything in life, practical thoughts are essential. Conversely, distracting, useless, repetitive thoughts only serve to scatter your focus and prevent you from entering the zone.

Practical thoughts will help an athlete to plan for an upcoming event, put together a training schedule, and organize any logistics such as booking travel or accommodation. Prior to a game an athlete might channel his thoughts into rehearsing the perfect shot or visualizing winning the race with confidence. Mental imagery like this can serve as useful preparation and is a positive way to use your thoughts.

Distracting thoughts, however, will be habitually negative and irrational, often so automatic that the individual is unaware of them. This type of thinking creates tension, draining you of energy and distracting your game. As the Irish writer Oscar Wilde once said, "Thinking is the most unhealthy thing in the world, and people die of it just as they die of any other disease."

Thinking is a habit. We have been taught how to think by the influences of society, education, and our friends and family. From a young age we are conditioned to believe that thinking is a good thing—how many times have you been told to "think about it" or "think it over"? Developing the mind into a sharp analytical tool is obviously very useful for achieving in the world of academics and other livelihoods, and while it is important that we learn how to think, it is equally important that we learn how to stop thinking. Just as a fire provides comfort and warmth from the cold, it can also be a destructive force when out of control. The mind is a similarly double-edged sword.

At this point you might be thinking, *What is he talking about? I'm in control of my own mind!* But are you really? How easily can you turn off your thoughts? I would like you to try a simple exercise: Stop thinking and observe how long it takes before the first thoughts enter. Is it 5 or 10 seconds, perhaps?

The degree to which you are in control of your mind depends on how long you are able to remain free from thought. The longer you can remain effortlessly free from thought, the better your powers of concentration and focus. Most people are only able to clear their mind for a few seconds at most. It is likely that your mind has more control over you than you may have imagined. The good news is that there is much to gain from taking control of your thought processes. Learning to quiet your mind is easy to do with a little focus and practice. Think of it as a challenge, just like any other new form of training, but one that will inevitably improve your health and sports performance.

The first step to escape the trappings of a busy mind and take control of your thought processes is to become aware of the thoughts inside your head. Seldom do we observe our minds. Seldom are we aware of our thought activity and the effect it has on mood, tension, and performance. Bring your thoughts to the forefront of your mind. When you first observe the activity in your head, you might find that your thoughts become amplified. This is simply because they are under your scrutiny rather than running unhindered in the background. You may also discover that the same repetitive thoughts have been running through your mind for quite some time, possibly for years. This is also normal, so don't be critical of yourself. Observing the mind is a most positive activity, allowing you to realize just how active it is. This awareness is the first step in allowing you to break free from the trappings of the mind and improve your concentration and focus.

When you take time out to observe your thoughts, you will realize just how often you are stuck inside your head. There is no need to analyze or judge what is going through your mind—doing so will only pile on more thoughts. The mind is never calmed by further questioning, and thinking too much is the very habit that we want to address. In order to step out of this constant circle of thought—to get out of your head and into life—you must learn to tame your mind.

By simply making the commitment to pay attention to your thoughts periodically throughout the day, you can take control of the most important tool you possess—the power of your mind. The clarity of your mind plays a significant role in determining your quality of life: A still mind brings benefit to your sleep patterns, mood, and health, while a mind full of incessant thoughts and distractions will blunt your ability to reach your full potential.

It is especially important to observe your thoughts when an undercurrent of negative thought takes over. Don't let yourself be overwhelmed by self-doubt and worry. The mind is unable to differentiate between an imagined and a real event; to the body, both are the same. If you are experiencing pregame nerves or overthinking the decision of a coach or worrying about being dropped from the team or anxious that you won't be able to complete that charity fun run, your body will react as if the event has already taken place. As soon as you become aware of an anxious train of thought, take notice of its effect on your body. Does your head or stomach feel tense? Is your breathing getting faster? Are negative and repetitive thoughts causing you to feel nauseated? How you think determines how you feel, and how you feel feeds back into how you think. A feeding frenzy of negative thought and emotion will inevitably result in poor performance and health.

Each time you find your head full of internal chattering, ask yourself if all this thinking and analysis is actually helpful. Is it getting you anywhere? Is it helping to solve the problem? If you continue thinking in this way, will it do anything to address the situation, or is it just a never-ending merry-go-round of habitual anxiety? Asking these questions provides an insight into the nature of repetitive thought and the way it affects your life. The realization that these negative thoughts serve no purpose to you can offer the impetus you need to escape from the clutches of your mind.

Be patient with your observation. At first, your thoughts may continue unabated as you start to notice the tension created within your body—your heart beats faster, butterflies flutter in your stomach, and your mind refuses to cease its activity. In time, and with practice, you will learn to still your mind and take back control. Observe your thoughts as many times as possible throughout your day, especially when faced with a challenge. There will be times when you can quiet your mind effortlessly; at other times the buildup of emotions may make it more difficult. Either way, observe how negative thinking affects your body and ask yourself if these types of thoughts are useful to you. Simply asking yourself this question will bring awareness to your life and allow you to know yourself better.

In time, if you commit to observing your thoughts and the effect they have on your body, you will find you spend less energy on fruitless thinking. Your mind will be clearer, your body more relaxed, and life will become easier. You will notice more of what takes place around you and miss less.

Repetitive, regurgitating thought is extremely draining, causing stress, tiredness, and headaches. The less headspace you give to negative thinking, the more room you have in your life for positivity and improvement.

Consider the following example:

Michael is driving his car to an important training session, well aware that he is likely to be late. As he projects his mind into the future, he imagines and rehearses the reaction from his coach and teammates. He cannot stop these thoughts circling around his head, worrying about the outcome, getting frustrated with traffic, and generally winding himself up. His body feels tense, a headache begins to take hold, and he probably starts to drive faster than he should.

Alan is also driving on his way to a training session. It is likely that he too will be late. Knowing that he is doing the best he can to get there on time and that no amount of worrying is going to get him there any faster, Alan brings his attention to his breath, observing the flow of air as it enters and leaves his body. Every now and again the thought of being late enters Alan's mind. As he notices the thoughts, he asks himself whether these thoughts are helpful. Knowing only too well that anxiety will only serve to make him tense and distract him, he takes his attention back to his breath, keeping himself calm and relaxed.

Now imagine both Michael and Alan are cut off by another car, forcing each of them to apply the brakes. How do you think each one reacts to the situation? Most likely, Michael is engulfed in rage and sounds the horn angrily. Alan, on the other hand, chooses not to react and does not get sucked into the situation.

In this example, the two drivers were presented with the same challenge, but while one immediately reacted to the challenge, the other responded by choosing to let it go. There is no doubt that a stressed or anxious person reacts more strongly to a difficult event than a relatively calm person. Keeping a quiet mind enables you to consider a situation more objectively and choose how to respond rather than being consumed by the moment. Alan had the choice to respond or not. Michael's instincts took over before he could make a conscious decision.

In addition to observing your thoughts, it is just as vital to be able to quiet your mind. Together, these skills will magnify your ability to enter the zone during sports. When you first practice taming your mind it is normal to find distracting thoughts entering your head every few seconds. In fact, you should expect them to appear. Do not be disheartened—this bombardment of thoughts has built up through years of conditioning and will take time to strip away. Layer upon layer of thoughts have been added by every influence in your life: education, religion, society, relationships, and work. The mind has simply developed a bad habit; it knows how to think but is unable to stop thinking.

When you first begin practicing meditation, try not to become frustrated when the mind wanders, because it inevitably will. Many people abandon meditation too quickly when they find it isn't an instant fix. You may feel discouraged if your thoughts don't switch off as quickly as you had hoped, but try to view the practice as just that: practice, not as a fixed outcome or goal.

Your intention while meditating should be to make a dedicated effort to become aware of your own mind, to observe your thoughts, and to practice simply being still and present. Thoughts will enter. Thoughts will go. That is the nature of the human mind. Each time you notice your mind wandering, all you need to do is gently bring your attention back to your breath or bring your awareness to your inner body.

All throughout my time at school, college, and my early years of work, my mind was extremely active. I believed that thinking was a positive thing but had no way of differentiating practical thoughts from repetitive, unnecessary, or negative ways of thinking. Most of the time I lived stuck in my head, on automatic pilot, without realizing just how constantly thoughts were streaming through my mind. Having such an active mind diminished my concentration—to achieve good grades it was necessary for me to spend many hours studying, and I found it difficult to retain information as there simply wasn't any room left in my head. For one university exam I spent three months studying in the Berkeley Library at Trinity College Dublin. An hour before the examination I headed there again with all my notes for a last-minute recap and was joined by my friend Terry, who hadn't prepared at all. While the rest of us had been studying business, Terry had devoted all of his time setting up a tax reclamation service. He borrowed my notes and studied them for a mere 15 minutes. As I observed his unwavering and effortless concentration, I could not help but notice that Terry had something I did not. My concentration at that time was so poor that I was easily distracted, and the little attention I had was forced. When the exam results came back, I was somewhat scandalized to discover that Terry had secured the very same mark as me. What he achieved in 15 minutes took me three months. This is exactly the difference between a productive, focused, and concentrated mind and one that is out of control.

To get through university I continued my long hours of studying, experiencing stress, fatigue, and severe respiratory problems in the process, completely unaware that my mind was working against me. Twenty years later, it comes as no surprise that Terry's company employs one thousand people, winning countless awards in business and customer service.

A skittering mind, jumping from thought to thought, is a leech to productivity, creative endeavor, and quality of life. Having a focused mind is probably the greatest asset in every walk of life, whatever your occupation or lifestyle.

After university I secured a job in middle management with a U.S.-based car rental company. We were indoctrinated into the philosophy of the company, molded to perform, and encouraged to give the best years of our lives for profit. We were told exactly how to answer the phones, how to deliver spiels to potential customers, and how to upsell collision damage waivers. There were targets to be met, employees to be managed, and sales calls to be made. Each Monday morning I dreaded the thought of going in to work. I was a living wreck at age twenty-four. My mind never stopped, my stress levels soared, and the more I thought about leaving the job, the more my mind kept me trapped in it.

Just as I was reaching the breaking point I heard about a course in personal development and jumped at the chance to enroll. During the course the instructor spoke about the importance of a still mind, and guided us through a simple meditation. After the very first session I noticed my perception was suddenly much clearer and lighter. The tension had left my head, my mind was quieter, and for the first time I caught a glimpse of stillness. As I walked home, it was as if my thoughts had been put to one side, allowing me to focus my complete attention on the sights, sounds, and smells around me. I had walked down Grafton Street in Dublin many times before, but never had I actually *been* there. On previous visits, my attention had been completely stuck in my head. I would walk from one end of the street to the other without remembering any aspect of the journey. It is difficult to relate to life, or be part of life, when all your attention is stuck in your mind.

The next morning I found my mind swamped once again with internal chatter, but the unforgettable experience from the day before remained with me. This epiphany occurred in the late 1990s, and in the months that followed I made a dedicated effort to bring stillness of the mind into my life. I had many ups and downs on my initial journey to quieting my mind—there were days when my mind was out of control and I felt like I was getting nowhere—but I now regard that time spent in quiet solitude as my most productive ever.

We are conditioned to believe that in order to be productive and successful we must be constantly doing something. This belief, which forms the basis of modern society, is quite insane. We are not human *doings;* we are human *beings*. During my workshops, students are often astonished to hear that if I were given the choice between my degree—which I worked so hard for—and learning to reduce my thought activity, I would choose the latter without hesitation.

I do my best to make my life a meditation, and I would estimate that my thought activity has reduced by around 50 percent since I first began practicing stilling my mind. Now, my thoughts are more practical: I set my goals, decide on a course of action, and set out to achieve them. Because there is more space between my thoughts, there is room for creative ideas and solutions to enter. I bring my mind into stillness many times throughout the day by focusing on my breath or taking my attention into my inner body. Of course, negative thoughts do sometimes appear; I still get upset and angry from time to time, and I have no problem confronting and challenging another person if required, but this is part and parcel of being alive. In nature, animals will sometimes fight, but after the confrontation they will go their separate ways, living life instead of dwelling on the event for hours on end. Nature moves simultaneously with time. More often than not, we humans spend much of our lives either stuck in the past or anxiously trying to get to the future. How can we possibly expect to utilize the full power of our brains if all our attention is consumed by habitual thought?

Since I learned to still my mind, one of the main differences to my life is that when faced with a challenge I feel less daunted and am able to recover from a setback much more quickly. By waking up to what takes place in my mind, I have a greater choice over whether I continue with the stream of thought or step out of it. In my early twenties, I never realized that I had this choice. Until I understood that I was the prisoner of my mind, I was unable to differentiate between my all-consuming thoughts and who I really was. By drastically reducing nonsense thoughts, my mind is now free to concentrate on anything of my choosing. At age forty-one, my concentration, energy, focus, and happiness have increased tenfold compared to when I was sixteen—and all I have done is learn how to stop thinking.

My life was completely transformed by three simple techniques: breathing lightly, merging with my inner body, and bringing my attention into the present moment. Each of these practices has been instrumental in improving my quality of life, reducing unnecessary thought activity, helping to access intuitive intelligence, and enabling creativity in my work. These powerful exercises are straightforward, quick to learn, and can be easily incorporated into your way of life. Practice the following techniques regularly to begin reclaiming your mind and clearing your head.

Follow the Breath to Enter the Zone

This sports meditation is based on the exercise Breathe Light to Breathe Right, which was introduced on page 74. The objective is to follow the pattern of your breath and bring a feeling of relaxation throughout your body in order to quiet the mind. Following the breath involves observing the cycle of each inhalation and exhalation, and is a simple and useful method of internalizing your focus while shutting out any unnecessary thoughts.

When first beginning your meditation practice, try to choose a place where you will not be distracted. Sitting in an upright position helps you to focus, while closing your eyes helps to direct your attention inward. With experience, observing the breath and bringing your attention into the inner body can be practiced during any situation or activity.

Following the breath is an activity innate to humans and has been practiced since time immemorial. The breath is the bridge between the mind and the body. In order to enter the zone during any type of sports competition, the body and mind must be merged together so that you can become one with the game.

The breath can be felt entering and leaving the body at four distinct points. The first is the area just inside the nose, the second is the area at the back of the throat, the third is movement of the chest, and the fourth is movement of the abdomen. You may find it easier to focus on one point more than others—experiment as you practice to see which feels most natural for you.

To begin, place one hand on your chest and one hand on your abdomen, just above your navel. Follow your breath as it enters the nostrils or passes down the throat. Concentrate on the area you are breathing into—are you using your chest or your abdomen to inhale? Do you feel your chest rise and fall with each breath, or do you feel your abdomen move gently in and out? Do not change your breathing, simply observe it. At first, you might find your mind wandering—don't worry if it does but simply bring your attention gently back to the breath.

As you observe your breathing, continue to relax your inner body. Using mental encouragement, silently tell the muscles of your chest and abdomen to relax. As you feel your body soften, gently slow your breath. There is no need to deliberately interfere with the breathing muscles by tensing them or restricting your breathing. Instead, simply allow your breathing to become quiet and soft, using mental commands to relax your body.

The objective here is to reduce your breathing to the point where you create a mild to moderate hunger for air. This need for air should be distinct but not so strong that your breathing accelerates, or your abdomen muscles contract, or the rhythm of your breath becomes chaotic. If you do find that your breathing becomes disturbed by the hunger for air, then take a break from the exercise for about 15 seconds and resume when your breathing is back to normal. Try this breathing practice for 10 minutes or so.

Breathing exercises of all kinds are helpful for taking attention from the mind and into the present moment. However, the creation and maintenance of a tolerable air shortage over the course of this exercise can be instrumental in further slowing down the activity of the mind. There is nothing like the feeling of an air shortage for anchoring attention to the breath. An additional benefit to breathing lightly is the activation of the body's relaxation response, indicated by increased production of watery saliva in the mouth. The technique and effects of reduced breathing are explored in more detail in chapter 4.

When I first started to use this method of observing the breath during meditation, I sometimes became frustrated to find my mind wandering. This is natural. In fact, you should expect your mind to wander at first. With practice, your mind will wander less and less, the gap between thoughts will increase, and you will feel happier and more alert.

By practicing meditation regularly, you will find that your attention no longer spends so much time in your mind, separated from life, dwelling on things you cannot change and things that have not happened yet. Over the past fifteen years I have completed several periods of "noble silence"—a series of ten consecutive days when I wake up at five in the morning and meditate until it is time to retire to bed at eight o'clock at night. Car keys, computer, phone, and wallet are put away. Noble silence involves a silence of the tongue and mind—no talking or thinking. At the end of the ten days, my mind is razor-sharp, calm, alert, and focused.

For those starting out in meditation, even a short period of 10 minutes observing the breath can bring about great changes in your life. Spend two weeks observing your breath as often as you can throughout the day. More important than the *length* of time you observe your breath during one sitting is the number of times you bring attention to your breathing throughout the day. See and feel the difference that this practice makes to your focus and concentration during sports and in your everyday life.

Connect with the Inner Body to Enter the Zone

Focusing on oneself involves taking attention out of the mind and merging with the miracle that is the human body. There is an intelligence that operates within our bodies, far greater than that within the mind. Every moment, thousands of functions automatically take place within the body without any direction from the mind. The intelligence of the mind is but a mere fraction of the natural intelligence that resides within the body. Consider the many vital automatic functions of the body, such as breathing, the beating of the heart, and the process of digestion—conscious thoughts are not required for these complex and tireless processes, but imagine if just one of these activities was the responsibility of the human mind. Even if the mind was only responsible for directing each breath—a relatively straightforward procedure—it is unlikely that we would last for more than an hour. The human body is a miracle and operates on such a vast innate intelligence that it is unlikely human beings will ever be able to produce comparable technology. Despite man's best efforts to develop a robot to parallel the human body, prototypes are awkward and capable of handling only a small number of functions. The natural intelligence that resides in the human body is incredible, and we all have access to its capabilities so long as we do not allow the mind to overshadow it. Simply by taking your attention from incessant thinking and directing your senses into the inner body, you will be able to draw from the stillness and intelligence residing there.

In the Western world we rarely focus on the inner body unless something is wrong. Seldom do we pay attention to the body when it is free from pain. Seldom do we truly experience our body as alive and feel the vibration of energy that resonates there. The body is your connection with this energy. If you have never directed your attention to your inner body before, start off slowly. If you have learned to follow and reduce your breath, you will find this relatively easy to do.

Close your eyes and bring your attention to one of your hands. Direct your focus to the hand and feel it from within. With your eyes closed, feel the inner sensations of your hand. You may begin to notice the temperature of the air against your skin, or you may feel subtle inner bodily sensations. Stay with these sensations for a little while and quietly observe them. When you are focused on the inner bodily sensations of your hand, move your attention to include your arm. Now feel the inner bodily sensation of your hand and arm together. Do not analyze or think about it, just feel it.

Next, bring attention to your chest and focus from within for a minute or so. You may feel the texture of your clothes against your skin, or you may feel the heat emanating from your chest. After a minute or so, bring attention to your stomach and check for any tension. If your stomach is tense, imagine it gradually relaxing. Feel the area around your abdomen gently softening as the tension dissolves away. The more active your mind, the greater the tendency that your stomach will be in a knot. Relax this area using your imagination and allow any tension to disappear.

Now feel the energy field of both hands, both arms, your chest, and your abdomen at the same time. Keep your attention there. As long as your attention is on your body, it is not on your mind. The incessant thought activity of your mind will gradually slow down as your focus remains on your inner body.

With a little practice, you will be able to focus your attention on your inner body from head to toe. Dispersing your attention throughout the inner body is especially helpful during physical exercise and competition, and this exercise can help you to enter the zone at will.

Live in the Now to Enter the Zone

By following the breath and bringing attention to the inner body, we are able to bring our attention to the present moment.

The present moment is the only time in which life truly unfolds. You cannot re-live your life in the past, nor can you live your life in the future. When the future arrives, it is the now. Be here fully. Do not spend your entire life, as most of humanity, with all of your attention stuck in your head. How can you relate to the reality of life if you are too busy focusing on memories, worries, and what-ifs?

A simple practice to bring your attention to the present moment is to merge with your surroundings. We connect to our surroundings in a physical way, through the five senses of sight, sound, touch, taste, and smell—not by intellectual perceptions. Put aside the habit of analyzing, judging, labeling, and comparing everything you see. Instead, bring a gentle focus to your surroundings without the usual running commentary. Take your attention out of your thoughts and look around. *Really* look, as if you are seeing things for the very first time. As you look, begin to listen to the sounds rising and falling around you. As you look and listen, feel the weight of your body, whether you are standing, sitting, or lying. Feel the warmth or coldness of the air on your face. Feel the clothes on your back. Bring in your other senses of smell and taste. Now you are free from thought. Now you are free from internal noise and distractions. You are like a child, seeing everything for the first time. It is that simple.

Live Your Daily Life in the Zone

Quieting the mind should not just be limited to the time spent in formal meditation. Instead, your whole life should be a meditation. Each day, as you go about your daily affairs, bring your attention to your breathing and your inner body. As you watch TV, do not surrender all of your attention to the program, but immerse yourself in your inner body. As you walk, jog, or run, follow the natural rhythm of your breath and disperse your attention throughout your body. Scan your body for any tension that may be residing there, and bring a gentle feeling of release to tense areas to encourage relaxation. Tension of muscle groups during sports is counterproductive and consumes energy—learn to recognize areas of tension in your body and practice melting them away with the power of your mind.

To observe complete relaxation of all muscles groups, watch a video of a cheetah gracefully running at full throttle. The leg muscles look to be completely relaxed and floppy as the animal moves effortlessly between strides. The cheetah is able to devote all of its energy to its speed through complete relaxation as it runs. Practice running with your entire body instead of just your head. Imagine running without a head. Be headless for the duration of your run. Run with every cell in the remainder of your body. Merge with the movement and become one with it. Bringing a feeling of relaxation throughout your body while you exercise allows you to go with the flow and enter the zone. The greater the proportion of your daily life that is spent unhindered by thought, the easier you will find entering the zone during competition.

A Concentrated and Undivided Mind

Imagine a pool player who wins ten games in a row. During each game, he is in the flow, pocketing balls easily while positioning the cue ball perfectly for the next shot. His game is effortless and spontaneous.

Based on his success, the player decides to place a wager of five dollars that he will win the next game. No longer is he doing the activity for the experience or enjoyment of it. Now he has an ulterior motive, and his mind is divided between the wager and playing the game. Only part of his attention remains on the activity. He has lost focus, and he loses.

Ronan O'Gara is a former rugby union player who played for Ireland and Munster. Considered by many to be one of the best fly-halfs of his generation, he won four Triple Crowns with Ireland and two Heineken Cups with Munster. He also scored more tries for Ireland than any other fly-half in history and is the third most capped player in rugby union history.

With such impressive repeated successes, one would expect O'Gara to have been full of self-confidence, approaching each game in his stride. However, in a compelling interview with Irish national television, he described how he wished he could have "laughed and enjoyed it more," especially in the lead-up to games where he would be "puking, questioning everything, not sleeping, feeling low, and going for walks" to try to quell his negative inner commentary. O'Gara is not alone in this experience of pregame anxiety, and no doubt his feelings are shared by many professional athletes who, in their sense of pride and dedication to their team and country, place immense pressure upon themselves.

This is the side of professional sports that is seldom talked about. It is brushed under the carpet and hidden from view while the world gets to see only the outward displays of power, strength, speed, and skill. After playing for over a decade at the professional level, Ronan O'Gara began to settle down only when he knew that his retirement from professional sports was imminent: "It was only in the last eighteen months that I said I'm going to start enjoying myself for whatever limited time is left."

Why is so much anxiety and tension created prior to a game? Prematch days for many athletes might involve hundreds of "what if" thoughts: *What if I don't feel right on the day? What if I get injured? What if I make a mistake? What if I mark the wrong guy? What if I'm not selected? What if I perform badly and get dropped from the team?* As imagination runs riot, anticipation of all the possible things that could happen during the game builds so high that nervous tension and anxiety take over to the point of distraction. In reality, there are so many variables involved that it is impossible for the mind to predict what is going to happen. Not only that, but the mind is capable of creating an imaginary drama with outcomes far worse than what might actually unfold. It is essential to understand that in any situation there are many factors outside of your control, and despite your best efforts you are but a part of the larger whole. Ruminating on anything outside of our control is simply a distraction and a waste of energy. Having passion for what you do and the desire to perform to the best of your ability are all that matters. Everything else is in the hands of the universe, so there really is no point in worrying about it. Before a game or event, observe negative or recurrent thought patterns and make a conscious decision to step away from thinking about things that you have no control over.

I may not be an athlete, but I can closely relate to the process of thinking so much that your thoughts consume you. As a sixteen-year-old boy from a Catholic background, educated in an all-boys school, I was particularly shy when it came to talking to girls. From time to time, a very attractive blond girl would take the same bus as me to school. I longed for a conversation with this girl, dreaming about what I would say, nervous that she would rebuke my advances. On one particular day she sat right next to me, and while my heart thumped out of my chest, no words could come out of my mouth. I was completely tongue-tied in a sea of doubt and sat the whole ten-mile journey in silence. I had built up the moment into something so big, so monumental, that it scared the life out of me. In hindsight, all I really wanted to do was strike up a conversation, to say hello and ask her how school was going, to ask whom she hung out with, and maybe talk about our favorite music. It was all very innocent, but overthinking created a task of gigantic proportions, almost as if I were planning to ask the girl to marry me.

With experience and confidence we learn how to deal better with situations like this, but by persisting in overthinking every action, we add unnecessary hurdles to the path to success. There is no doubt that had I thought less about striking up a conversation with the girl, it would have happened naturally and easily when the situation arose.

Although a certain amount of stress helps to keep us focused, too much thinking, anxiety, and apprehension about what can go wrong can cause a loss of concentration on the game. The night before a big event you may find yourself lying in bed ruminating on every possible scenario, while in reality a deep sleep is really what you need to secure a good performance the following day. There is a time for planning, and it is more likely to be effective when you are able to focus on it productively. Pregame warm-ups, drills, rehearsals, and discussions of tactics can help to reduce stress and iron out uncertainties, but lying awake worrying about what might go wrong will only cause more self-doubt and potentially ruin your concentration on the day.

It is important to consistently monitor the effectiveness of your thinking. When you begin to notice the same thoughts creeping into your head for the umpteenth time, ask yourself whether these thoughts are actually serving a purpose. Are they helpful for devising a strategy or addressing the situation? Or are they keeping you on a merry-go-round of insanity? Questioning the usefulness of your thinking allows you to determine which thoughts are useful and which are negative and repetitive. Even though you may not be able to prevent these thoughts from occurring, you can learn to reduce them by quieting your mind and bringing your attention to the breath in between bouts of over-thinking. Later, if you feel the need to think a little more about whatever is bothering you, spend a few minutes indulging your thoughts before merging back with the quietness of the breath. Alternating between thinking and quietness creates space between thoughts to allow fresh ideas to surface. The thoughts that emerge after the silence of meditation can be very powerful, creative, and intuitive, and may naturally provide you with the solution to your worries. The same process of course applies to daily life.

Practicing meditation and achieving a quiet mind are extremely valuable techniques for athletes, or anyone who suffers from anxiety, enabling you to reduce your inner commentary and focus on what's important. To reduce prematch apprehension and nerves, you will need to approach the game from a different mind-set. Participate in your sport solely for the experience, because you enjoy it. When your only wish is to experience the game or the individual shot, swing, or race, there is no ulterior motive. This doesn't mean that you will simply be going through the motions—no, not at all. When you are playing or running or shooting or cycling solely for the experience, your mind will be undivided and a state of intense alertness will ensue. In this state, you will be at the top of your game.

At first, make a commitment to really experience your sport during training sessions, where there is nothing to lose. Fully experience your sport by dispersing your attention throughout your entire body and getting in touch with your senses. In time, you will feel comfortable "experiencing" your sport regardless of the level of competition, allowing muscle memory to ensure a smooth and effortless performance.

Make it a priority to occupy your body with your full attention and experience the aliveness of your entire body. During a race, follow your instincts and your natural rhythm, letting muscle memory decide on the course of action: how far to stay back, when to overtake, what move to make next. Before taking a penalty kick or another game-changing action, focus on your breathing and use the breath as an anchor to your inner body.

Whenever you find your mind ruminating about an upcoming event, immediately bring your attention to your breath or inner body. If you are trying too hard to win, psyching yourself out about the outcome, worrying about your competitors and past failures, or spending too much time analyzing every potential move, your mind will be divided and you will lose focus.

Improve Brain Oxygenation

It is perfectly normal for athletes to be nervous before competition. But while a little nervousness keeps us alert, too much may induce hyperventilation, which reduces oxygenation of the brain. You might not be sitting for an exam, but without a doubt, mental alertness, concentration, and normal cognitive functioning are prerequisites for good performance.

In addition to following the breath and occupying your inner body with your attention, the following Breathing Recovery Exercise is very helpful in calming the mind during the days and nights leading up to competition. When stressed, hold your breath! It is also helpful for recovering from physical exercise and increasing your BOLT score. Having a high BOLT score will help negate the effects of nervousness.

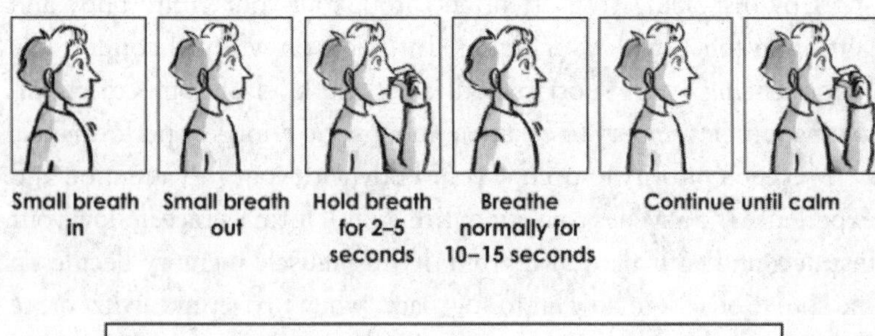

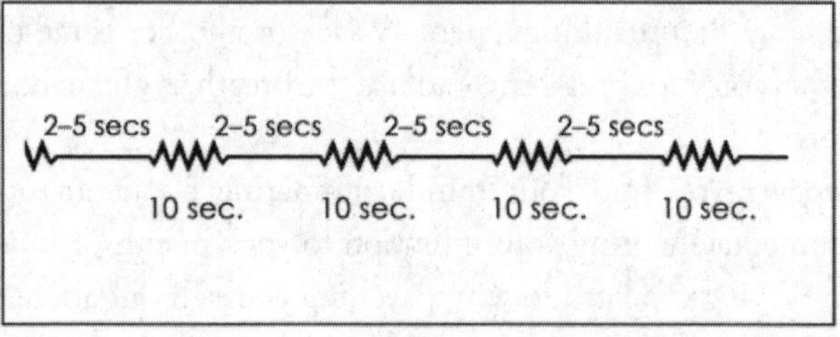

Perform a series of small breath holds following these instructions:

- Take a small, silent breath in and out through your nose.
- Hold your breath for 2 to 5 seconds.
- After each breath hold, breathe normally for around 10 seconds. Do not interfere with your breathing.
- Continue to do a small breath hold followed by normal breathing for around 10 seconds.
- Practice this exercise for at least 15 minutes.

Having a BOLT score of less than 20 seconds during rest indicates chronic overbreathing. To attain optimum performance in terms of breathing efficiency, cardiovascular health, and delivery of oxygen to tissues and organs, a BOLT score of at least 40 seconds is necessary. Researchers have observed that hyperventilation significantly affects mental performance as well as physical capabilities. A study investigating unexplainable aircraft accidents tested the abilities of jet fighter pilots to use coordination apparatus after a short period of breathing too much. The results showed that mental performance deteriorated by 15 to 30 percent when the concentration of carbon dioxide in the blood was significantly reduced. Another study found that when hyperventilation reduces arterial concentration of carbon dioxide, physiological changes occur in the brain, causing dizziness and concentration problems. Researchers discovered that reduced levels of carbon dioxide detrimentally affected performance that required attention, causing progressively slower reaction times and an increase in errors.

The symptoms of hyperventilation and anxiety are similar and have been found to be linked in some cases. A study from the department of psychology and statistics at the University at Albany in New York found that students with high anxiety had lower levels of carbon dioxide and a faster respiration frequency than students with low anxiety. The results of this study are not surprising when you consider the effects of hyperventilation: dizziness, headaches, chest pains, and light-headedness. Is it the anxiety that is causing hyperventilation, or is it hyperventilation that is causing the anxiety? As we already know, hyperventilation reduces the concentration of carbon dioxide in the blood. This leads to a narrowing of blood vessels and reduced delivery of oxygen to the brain. An oxygen-deprived brain is more excitable and agitated, and as it floods with self-generated thoughts, anxiety kicks in. One contributes to the other, creating a vicious and self-perpetuating cycle.

I vividly remember attending one of my final exams at Trinity College in Dublin. In an effort to relax, I took a brief walk before the exam, during which I took several big breaths through my mouth. I was already a heavy breather and the additional big breaths brought on light-headedness and dizziness. Little did I know that my anxiety and deliberate attempts to take deep "calming" breaths were actually depriving my brain of oxygen—not an ideal recipe when total alertness and concentration are required. Often this is an unconscious activity that athletes do without even realizing it. However, the vast majority of athletes also feel that taking big breaths is beneficial. The belief is there, but they are not always aware of practicing it. Overbreathing is never the answer to improving performance, and adequate oxygenation of the brain is required if you are to fulfill your true potential.

Quality Sleep for Performance

To help maintain a calm and concentrated mind, you must experience good-quality sleep at all times, particularly during the period leading up to an exam, performance, or competition. Having a BOLT score of less than 20 seconds and breathing through the mouth during sleep may result in many of the following symptoms:

- Snoring
- Sleep apnea (holding your breath many times throughout the night)
- Disrupted sleep
- Insomnia
- A racing mind
- Nightmares
- Sweating
- Needing to use the bathroom at around 5 or 6 A.M.
- A dry mouth upon waking

- Brain fog upon waking
- Fatigue first thing in the morning
- Fatigue during the day
- Poor concentration
- Upper or lower respiratory complaints

Mouth breathing during sleep creates a considerable loss of carbon dioxide while also bypassing the benefits of nitric oxide, culminating in a reduction of morning BOLT score. The solution to better breathing at night is to Breathe Light to Breathe Right (see page 74) during the day and especially before sleep. To reduce and eventually eliminate breathing through the mouth at night, follow these guidelines:

- Avoid eating within the 2 hours before sleep, as the process of digestion increases breathing.
- Keep your bedroom cool and airy (but not cold). A hot and stuffy room will only serve to increase breathing.
- Sleep on your front or left side; sleeping on your back is by far the worst position, as there is no restriction to your breathing.
- Ensure that your mouth is closed while you try to get to sleep (you can wear 1-inch Micropore tape across the lips as described in chapter 3 to help ensure this).
- A most important practice to discourage heavy breathing at night is to Breathe Light to Breathe Right for 15 to 20 minutes before going to sleep. This exercise is especially good for calming the mind and helping you to experience deep sleep during the nights leading up to competition.

PART III

The Secret of Health

CHAPTER 9

Rapid Weight Loss Without Dieting

For many weekend warriors, the motivation to exercise is to shed a few extra pounds for better health, a boost of self-confidence, and that feel-good factor. There is no doubt that exercise is a good route to improved health and weight loss, but it addresses only half of the story. Weight loss occurs only when the amount of calories we burn is greater than the amount we consume. In addition to keeping an eye on the pedometer, we also need to stand back from the table. This is where many people fail, ending up on a yo-yo diet of weight loss followed by weight gain in a never-ending cycle of frustration.

For over a decade I have witnessed hundreds of people achieve a safe method of appetite suppression leading to steady, effective weight loss using breath reduction techniques. Weight loss for these individuals varied from 2 to 6 pounds within just two weeks. In addition, people often found themselves to be eating more heathfully with less desire for processed food and more demand for water. What's more, this weight loss and change to better eating habits occurred easily and without effort. In many cases weight loss was actually a secondary benefit, as most participants were applying the breathing exercises to remedy asthma, anxiety, or snoring. The only instruction they were given with regard to their diet was to eat when hungry and stop when satisfied.

When Eamon entered his fifties, he tipped the scales at more than 260 pounds. Ireland was going through a rough time economically, and Eamon's business was not doing very well. The increased stress of trying to steer his business back on track led him to eat and drink more than ever before. Nearly every night he headed down to his local pub to meet with friends and drown his sorrows. Within two years, Eamon was diagnosed with high blood pressure and type 2 diabetes.

For years, Eamon had lived a comfortable life in good health. As a younger man he played sports and exercised regularly. Now he found himself caught in a vicious circle of feeling low and despondent, unable to focus and unable to get his business or health back on track. Following the untimely passing of a close friend, Eamon decided that he had enough and said to himself, "I've got to change my life." This was his wake-up call.

When Eamon contacted me to help reduce his stress levels, he was desperate to improve his situation: "My business has me totally stressed out, I'm not sleeping at night and can't even think straight." My main focus was to help him get back to good health by retraining his breathing. With an increase in energy and concentration, the rest would follow.

Eamon's starting BOLT score was just 8 seconds, and he displayed traits of someone dealing with stress and anxiety: taking large breaths from the upper chest and sighing regularly. Eamon's first steps were to learn how to breathe through his nose day and night, to take time out to relax and meditate, and to practice the Breathe Light to Breathe Right exercise. Stress was the main issue to be resolved, and learning how to still the mind was an essential part of addressing his other symptoms.

For people with diabetes and high blood pressure, it is important to go slowly and gently when practicing reduced breathing exercises, so as not to put additional pressure on the body. Reducing breathing commonly leads to a reduction of blood sugar levels, and while this is a good thing, decreasing levels too quickly is not. As Eamon's BOLT score increased, both his diabetes and high blood pressure medications were reduced accordingly by his doctor. If you have high blood pressure or diabetes, it's important to consult your medical professional before embarking on a reduced breathing program.

Eamon's regimen was as follows:

- Perform 10 minutes of Breathe Light to Breathe Right (page 74), 4 times per day, with 1 session just before bed and 1 just after waking.
- Stop and observe the breath for 1 to 2 minutes at various times throughout the day to further reduce overthinking.
- Tape the mouth closed before retiring to sleep to ensure nasal breathing at night.
- Practice the Breathing Recovery Exercise (page 91) whenever he felt anxious or stressed.
- Walk with the mouth closed for 30 minutes daily.
- Pay attention to appetite and eat only when hungry.
- Reduce alcohol consumption to no more than 2 servings of alcohol each night for the first week. From week 2, reduce drinking to every other night.

From the start, Eamon questioned whether these techniques were actually going to work. They flew in the face of everything he had been told to date—well-meaning stress counselors had encouraged him to take deep breaths—and reducing his breathing seemed like the opposite of what he needed.

During our first consultation, Eamon practiced the Breathing Recovery Exercise of holding his breath for 5 seconds, followed by normal breathing for 10 seconds. He continued this for about 5 minutes before taking a rest. Following the Breathing Recovery Exercise I asked Eamon to place his hands on his chest and abdomen, and to apply gentle pressure with his hands to slow down his breathing and create a light air shortage. He practiced Breathe Light to Breathe Right for 3 minutes. Despite experiencing a mild urge for air, within minutes he started to feel tension lifting from his head. I'm pretty sure this was the turning point for Eamon. Breathing light had improved his blood flow and body oxygenation after just a couple of minutes—assuring him that this was indeed the way forward.

From then on I saw Eamon each week for a month. His BOLT score made steady progress, and by the fourth week it had increased to 27 seconds. His sleep was far better, leaving him feeling more alert upon waking, and he experienced a tremendous improvement to his well-being. Eamon's high blood pressure and blood sugar levels also reduced, while his doctor observed his progress and altered his medication as required.

Another positive side effect from the reduced breathing for Eamon was a lessened need for food. His appetite was much lower than normal, and he gave up drinking completely on weekday evenings. With decreased stress levels, there was far less temptation for both alcohol and food. Eamon's friends and neighbors have commented on how well he looks now that his weight has been reduced by more than thirty-five pounds. I have met with him a few times since, and although his BOLT score progress has slowed slightly, he not only looks like a different person but feels it as well. Eamon's story stands out in my mind as he achieved a lot despite having a lot of distractions and difficulties going on. He was extremely dedicated and followed through with each task he was given. It is often the case that those who are the most unwell are the most determined to follow the program in order to get their health back. Suffering can be a very effective motivator, but it's even better to make just a few small effective changes before you reach the breaking point.

My objective for this chapter is to show you the relationship between breathing and food consumption, not to tell you what and what not to eat. Of course there are foods that are best eaten in moderation or eliminated entirely from the diet, and these are well documented in most books on health and diet. A much more useful approach is to look at the reasons why you might be stuck on a permanent yo-yo diet, or continue to struggle to lose weight, and the answer may be closer than you think.

We can live without food for weeks, without water for days, but without air for just a few minutes. In terms of importance for survival, breathing is at the top of the list, followed by water, with food in last place. Health professionals, athletes, and nonathletes alike pay far more attention to their food than their breathing, but what happens if we switch this focus around? Improve your BOLT score by 10 seconds and you will find your appetite changing. Improve your BOLT score to 40 seconds and your life will change.

The loss of appetite and resultant weight normalization from obtaining a higher BOLT score may be due to a combination of several factors, including one's blood pH shifting toward normal, the effects of simulated high-altitude training, or simply an increased feeling of relaxation helping to reduce emotional eating. In this section we will examine each of these elements to help explain why Oxygen Advantage exercises help reduce the appetite.

Overweight individuals tend to have poor breathing habits, such as chronic hyperventilation, frequent sighing, and breathing from the mouth and upper chest. Putting on a few extra pounds causes us to breathe more heavily, and not just during physical exercise—breathing volume is increased during rest as well. Based on my observations, there is a clear relationship between breathing volume and food consumption. The question is whether processed and acid-forming foods lead to the development of poor breathing habits, or might it be that poor breathing habits lead to cravings for processed and acid-forming foods? In my experience there is a feedback loop between breathing and weight gain, and this cycle must be broken if change is to occur.

The pH scale measures acidity and alkalinity, ranging from 1 to 14. On this scale, 1 is the most acidic, 14 the most alkaline, and 7 is neutral. As we saw in the first chapter, carbon dioxide plays a crucial role in the regulation of blood pH. Our bodies strive to maintain a state of balance known as homeostasis, which includes normal blood pressure, normal blood sugar, and normal blood pH within a narrow range of 7.35 and 7.45. This balance of chemicals is kept in check by the lungs and the kidneys. If blood pH drops below 7.35 it will become too acidic, causing breathing volume to increase as the lungs work to correct pH levels by offloading carbon dioxide (which is itself acidic). Over-acidity of the blood may occur when we eat processed and acid-forming foods, leading to heavier breathing and symptoms of bloating, lethargy, and weight gain.

Conversely, an individual who chronically overbreathes will expel too much carbon dioxide, increasing blood pH to alkaline levels above 7.45. One hypothesis for the relationship between overbreathing and weight gain is that the body craves processed and acid-forming foods in an effort to normalize blood pH. Correct breathing volume and a good diet work together to keep blood pH at a healthy balance.

Throughout evolution humans have adapted very well to coping with short-term stress. During short periods of stress, breathing volume temporarily increases as the fight-or-flight response is activated. Once the stress has dissipated, breathing volume will normalize once more, allowing carbon dioxide to accumulate and restore normal pH. However, when it comes to long-term stress, persistent overbreathing decreases carbon dioxide levels for extended periods of time, meaning blood pH is not given the opportunity to normalize.

Notable health practitioners around the world advise eating alkaline-forming foods such as fruit and vegetables and avoiding an excess of acid-forming foods such as animal protein, grains, and processed foods. And though the majority of us know what it means to eat healthily, the temptation of processed and sugary foods can sometimes be impossible to ignore. Are we just following the demands of our bodies, or is there a way to get rid of these urges for unhealthy food naturally?

Time and time again I have witnessed startling changes to the diets of my students who learn to practice reduced breathing, often without the need for deliberate avoidance or willpower. These individuals, when shown how to address their poor breathing habits and increase their BOLT score by at least 10 seconds, automatically find their diet changing to healthy foods over processed foods. It begs the question: Might breathing be the missing link in the majority of weight-loss programs?

Breaking the vicious cycle of acid-forming foods and increased breathing volume is certainly a factor in achieving weight loss and provoking a reduction in appetite, but there are other factors to consider when looking at the relationship between breathing and diet, such as the effects of simulated high-altitude training.

Since 1957, scientists have identified that animals lose weight when living at high altitude. Sherpas and others who reside permanently at high altitudes are also generally thinner than their sea-level counterparts. Based on this observation, there have been many studies that point to the benefits of living at high altitude as a way to reduce obesity. The reason for this sustained weight loss seems to coincide with the lack of appetite experienced at high altitude due to a reduced saturation of oxygen in the blood.

In tests with mice it was found that moderate exposure to lower oxygen saturation can reduce body weight and, just as important, the levels of blood sugar and blood cholesterol. Researchers concluded that this was due to increased synthesis of EPO by the kidneys. This discovery has a particular resonance with the Oxygen Advantage program, since breath holding has been shown to increase EPO by up to 24 percent.

Of course, living at high altitude is not feasible or even economically possible for most people, and ironically obesity is also a risk factor for developing acute mountain sickness. But you needn't climb a mountain to achieve sustained and effective weight loss. Reduced breathing exercises like those practiced in the Oxygen Advantage program provide a practical and accessible alternative to high-altitude training.

High-altitude training can be simulated by incorporating breath holding into physical exercise as well as practicing reduced breathing during rest. For individuals with a BOLT score of less than 10 seconds, or those who have any preexisting health concerns, I would recommend starting by becoming accustomed to breathing through the nose night and day. From there, you can practice the Breathe Light to Breathe Right exercise (see page 74) for 10 minutes, 4 times a day in order to reduce your breathing volume toward normal and increase your BOLT score. Even a gentle approach like this can be enough to kick-start a reduction in appetite and help toward achieving healthy weight-loss goals.

For those with a BOLT score of more than 20 seconds and who are relatively healthy, in addition to practicing the above exercise you can also begin to incorporate breath holding into your physical exercise to simulate high-altitude training, as described in chapter 7. Holding the breath during walking, jogging, or running to create a medium to strong hunger for air decreases the oxygen saturation of the blood to below 94 percent, which can lead to a suppression of appetite. I suggest that you add these exercises into your current training regimen to make the program easier to sustain over the long term.

Another reason simulated high-altitude training may contribute to weight loss is that walking or jogging with nasal breathing allows the body to work with oxygen (aerobically), while incorporating breath holds every minute or so makes the body work without oxygen (anaerobically). During an anaerobic state the body is forced to burn calories from fat stores in order to produce energy. Incorporating both aerobic and anaerobic workouts into your training program will lead to increased calorie burn and weight loss.

Finally, there are emotional and psychological factors to take into account when looking at the reasons behind increased appetite and weight gain. The relationship between stress and increased food intake is well documented, with food often providing a distraction or comfort to negate feelings of anger or loneliness, or financial or relationship issues. I imagine that most readers will have experienced, at one point or another, that their demand for food increases when they are bored, stressed, or feeling low. This is mostly an unconscious habit, just as smokers will light up a cigarette without making any conscious effort to do so. As if on autopilot, we find ourselves going to the refrigerator or cupboard to consume whatever is in sight, despite having no genuine sensations of hunger.

In a study conducted by the University of Minnesota, data was collected from more than twelve thousand individuals to determine the relationship between perceived stress and health behaviors. Results showed that high stress for both men and women was associated with smoking, a diet that was high in fat, and a reduction in exercise. As stress is known to increase food consumption, anything that helps to reduce the effects of stress can be useful as an aid to weight loss. Throughout this book I frequently discuss the importance of taking your attention from the mind and dispersing it throughout the body, focusing on the breath or the present moment.

When your attention is focused on the sensations of your inner body and your breath, it is impossible to maintain anxious, stressful, and distracting thoughts. These practices of meditation have been employed by human beings for thousands of years, and today many studies show the benefits of meditation as a means of helping with weight loss. Controlling stress and depression can be conducive to maintaining long-term weight loss, and while it is relatively easy to shed a few pounds with a burst of well-intentioned exercise and healthy eating, the Holy Grail is to reach and maintain your ideal weight—no one wants to spend their life struggling on a restrictive diet.

The exercise Breathe Light to Breathe Right on page 74 has been specially designed to help you bring your attention away from the mind and onto the breath. This exercise is performed while sitting or lying down, but a certain amount of focus and concentration is still required to follow the breath and gently allow it to reduce to create a tolerable air shortage. Observing your breath, slowing it down, and bringing a feeling of relaxation throughout the body is in itself a meditation. An additional benefit of practicing this exercise and nasal breathing is an improvement in body oxygenation, thereby reducing brain cell excitability and stress.

Meditation doesn't have to be about sitting in a corner in a lotus position saying "om." At first it will be necessary to find a quiet place to sit so that you can focus on observing your breath to practice, but in time this exercise can be brought into any aspect of your life. As described in chapter 8, make your life a meditation by taking your attention out of your head and into the current moment. It is impossible to experience life when we hide away from it by allowing the mind to be permanently occupied by worries and stress. You are not just a head. In fact, imagine you are headless. No matter what activity you partake in, bring your attention out of your head, into your body, and on to the activity. Become the activity. And you can apply this theory to your eating habits too.

Eating is a basic function that often receives very little conscious attention during our busy daily routines. It is common practice to stuff food into our mouths unconsciously, not even really tasting anything after the first bite. The next time you eat, observe how many times you actually noticed the texture, taste, and smell of your food. Did you eat most of your meal with your attention elsewhere? Or did you experience the sensory wonders of your food, bite by bite?

Forty-one-year-old Tessy considers herself a born worrier. As the eldest child, she often felt an overwhelming sense of responsibility toward her family and younger siblings; throughout childhood, her parents constantly reinforced the notion that Tessy's primary role was to show a good example to her younger brother and sister. At school she excelled at both sports and academic work and became focused on achieving top marks—anything other than an A grade was unacceptable. The few times she received a C grade, she and her parents would get very upset. On one such occasion her father went to the extent of taping her report card to the refrigerator door to remind everyone in the house that marks like this were not to be tolerated.

It sometimes seemed to Tessy that her parents were trying to realize their own ambitions through her, but that it was impossible for them to be fulfilled. Her brother and sister had far less pressure put upon them, were required to do less homework and fewer household chores, and enjoyed much more freedom. These inequalities became even more apparent during their teenage years when the younger ones were allowed to watch hours of television and go to parties while Tessy was expected to concentrate on her studies.

Tessy grew resentful of her parents. She hated the fact that there was one set of rules for her and another for her siblings. Through the molding by her parents, Tessy felt that she always had to be a people pleaser and a perfectionist, and if she failed to reach her high expectations, she would feel very low and critical of herself.

Things came to a head a year ago when Tessy's mother came to visit and stayed for a whole month. It was then that Tessy realized just how much her mother's high-strung personality made Tessy feel stressed. Although she had long been an adult, Tessy had never escaped the attentions of her overbearing mother, who advised her, smothered her, and always said she knew what was best for her. As her mother had grown older, Tessy chose to keep the peace and continued to bottle up her feelings rather than confronting her.

Unsurprisingly, Tessy's mother's visit was very stressful. In a bid to distract herself from the stress, Tessy found herself eating more and more. Whenever she felt that things were getting to be too much, she would cook a meal or head to the local diner to dissipate her feelings. She enjoyed the pleasure of cooking, tasting, and eating, and now the food was serving an additional purpose by helping her to cope.

Six weeks ago, I received a call from Tessy complaining that she was feeling increasingly short of breath. Feelings of dizziness while driving had made her feel very anxious about her health, suspecting that something more serious was going on, and she had begun to suffer from breathlessness, unable to take a satisfying breath. Recognizing the angst in Tessy's voice, I asked her to come in to see me the following day. She explained the situation that led to the changes in her breathing and the use of food as a crutch. She was conscious about having a "spare tire" around her waist, which wasn't helping with her self-esteem, and she was anxious about her breathing issues—the time had come to make a change.

We started by measuring Tessy's BOLT score and found it to be 10 seconds. Although she was predominantly a nasal breather, she used her upper chest to breathe and sighed regularly. I explained to her that optimal breathing should be difficult to see and not heard, with gentle movements from below the diaphragm. During stressed-out breathing, however, the opposite takes place, leading to an unnecessarily large breathing volume and all its associated problems. Tessy needed to learn how to use slow and gentle breathing to bring her body into relaxation.

First, I asked her to place one hand on her chest and one hand above her navel and to become aware of her breathing, to feel the air as it entered and left her body. When she was conscious of her breathing,

I asked her to breathe light and gently soften her breathing in order to create a tolerable need for air for a couple of minutes. Tessy felt uncomfortable with the feeling of air hunger, so I asked her to reduce the time spent on the exercise in order to gently condition her body to acclimatize to the sensation. She repeated 3 sets of 1.5 minutes each, with a small rest of 1 minute or so between them, and soon became accustomed to practicing the exercise.

To help quicken her progress, I decided to teach her a walking exercise with breath holds. This involved walking with her mouth closed for about 1 minute or so, followed by gently exhaling through her nose and pinching it with her fingers to hold her breath for 10 paces. Following each breath hold, Tessy continued to walk while breathing through her nose for 1 or 2 minutes, followed by exhaling through her nose and holding her breath. Tessy found this exercise much more comfortable and proceeded to increase the lengths of her breath holds to 15 and then 20 paces. Each time she advanced, I checked to ensure Tessy's breathing remained under control, and soon she was able to hold her breath for 30 paces, even feeling comfortable with the feeling of air shortage. Tessy explained that she found it easier to practice reduced breathing while walking than sitting still, as she knew that the air hunger was only temporary.

Tessy made such good progress that we decided to move onto abdominal breathing to change her breathing pattern from the upper chest to the diaphragm and address her chronic overbreathing. For this exercise I asked her to stand up, as an upright position provides an ideal posture for abdominal breathing, and gave her the following simple instructions:

- Breathe in—gently allow your abdomen to move outward.
- Breathe out—gently allow your abdomen to move inward.

By focusing her attention on her chest and abdomen, Tessy was able to switch from upper-chest to abdominal breathing effortlessly. The next step was to gently soften and slow down her breathing to produce a comfortable hunger for air. Tessy practiced lightening her breathing for 3 minutes before taking a break for 1 minute or so. After 3 sets of this exercise and a short rest, I measured Tessy's BOLT score again to find it had increased to 23 seconds. This drastic change to her BOLT score took place in a single one-and-a-half-hour session, and Tessy felt calmer, more alert, and in much greater control of her breathing. BOLT scores do not usually increase so quickly, but from time to time it is possible! However, as I explained to Tessy, her sudden high BOLT score would reduce again over the following few hours, but could be regained through the practice of the exercises she'd learned.

A few weeks after we met, Tessy reported that she had felt extremely thirsty on one of the days, almost as if her body was desperate to rehydrate itself with plain water after months of relying on fizzy drinks. Exhilarated by her progress, Tessy felt calmer, more energetic, and no longer turned to food for comfort, losing ten pounds simply through the practice of quiet, gentle, reduced breathing.

By practicing Oxygen Advantage exercises and making improvements to your breathing pattern, you will achieve a higher BOLT score and experience a reduction of your appetite. Listen to your body and observe what it is telling you. Get used to responding only to the feeling of genuine hunger rather than eating for boredom or in response to stress or depression. The next time you feel the urge to go to the refrigerator or eat a snack, ask yourself, "Do I really feel hungry?" By eating only when your body actually requires food, you will make the most of your natural suppression of appetite and find losing weight and maintaining a healthy diet achievable and simple. A detailed program for helping to achieve weight loss can be found on page 282.

CHAPTER 10

Reduce Physical Injury and Fatigue

My brother Lee and his wife, Marie, are in their early thirties and live with their two children in the town of Navan in Ireland. Their work life, family life, and social life all revolve around physical training and preparation for long-distance events, and every few weeks they participate in triathlons, marathons, and even ultramarathons. Most people outside their sporting circle regard them as exercise nuts, and every now and again, for a bit of devilment, another of my brothers—Dave, who does no physical exercise at all—sends Lee the most recent newspaper reports describing how too much exercise is bad for your health, increases the risk of developing different conditions, or causes premature death. There is nothing more pleasurable to a couch potato, it seems, than to extol the dangers of exercise to those who do it.

There is anecdotal evidence to suggest that athletes may be at risk of becoming seriously ill early in life or die younger than the rest of the population, despite being in peak physical condition. And while the benefits of exercising to maintain good health are well accepted by all health authorities, is there ever a time when exercise can simply be too much or too intensive?

To investigate the relationship between longevity and career success, Professor Richard Epstein and Catherine Epstein from the Kinghorn Cancer Center in Sydney, Australia, analyzed one thousand *New York Times* obituaries published between 2009 and 2011. Their findings revealed that sports players lived on average for 77.4 years, while longer lifespans could be found in the military, business, and politics, where individuals lived for 84.7 years, 83.3 years, and 82.1 years respectively. And while 77.4 years is a ripe old age, why should athletes live shorter lives than those who probably devote less time to their health and fitness?

In addition to professional athletes living shorter than their counterparts in the business world, there is much documented evidence that intense physical exercise increases oxidative stress that may contribute to premature aging, damage to the heart, and dementia.

Given that most health professionals encourage physical exercise for good health, in what circumstances might exercise be damaging? And, more important, what can we do to reap the benefits of physical activity without putting our health at risk? The key to answering these questions seems to lie in controlling the amount of stress put on the body during exercise—more specifically, oxidative stress, which results from too many free radicals washing about our system.

Free radicals are molecules generated by the breakdown of oxygen during metabolic activity. We all create a certain amount of free radicals through the very act of breathing, but normal levels do not pose a problem since the body's defense mechanism is able to neutralize the molecules with antioxidants such as glutathione, ubiquinone, flavonoids, and vitamins A, E, and C. But when our antioxidant defenses are overwhelmed by too many free radicals, cells can be damaged and our health adversely affected. This is what is known as oxidative stress.

Free radicals are highly reactive and attack other cells, causing damage to tissues and negatively affecting lipids, proteins, and DNA. During physical exercise we produce more free radicals than usual due to an increase in breathing and metabolism, which can lead to an imbalance between the production of free radicals and the antioxidants required to detoxify them, resulting in muscle weakness, fatigue, and overtraining. Investigations into physical training, regular aerobic exercise, marathon running, and extreme competitions have consistently found that antioxidant levels decrease after intense physical activity or extreme competition, while free radical production increases.

In a paper published in the *Journal of the American College of Nutrition,* Guillaume Machefer and colleagues investigated whether extreme running decreases blood antioxidant defense capacity. Blood samples were collected from six well-trained athletes participating in an ultramarathon called the Marathon of Sands. In what is considered to be one of the toughest foot races on earth, competitors run the equivalent of six regular marathons over six days in the Sahara Desert, during which they are required to carry their own food. Blood samples were taken from runners 72 hours after completion of the race, with researchers noting a "significant alteration of the blood antioxidant defense capacity" and concluding that "such extreme competition induced an imbalance between oxidant and antioxidant protection."

In an attempt to deal with this potentially detrimental imbalance between antioxidants and free radicals, athletes are often encouraged to take large regular doses of antioxidants. At first glance this might seem like sound advice, but studies exploring the use of dietary antioxidants to reduce oxidative stress and exercise-induced muscle injury have met with mixed results to date.

An alternative and totally natural method of protecting against the excessive buildup of free radicals is to supplement regular exercise with breath holding and to increase your BOLT score. This method is cheap, nontoxic, and less controversial than supplements, providing effective protection against oxidative stress. Breath holding after an exhalation causes a decrease in oxygen saturation, which triggers an increase in lactic acid. At the same time, carbon dioxide levels also increase, leading to a rise in the concentration of hydrogen ions, which further acidifies the blood. Repeated practice of breath-holding exercises offsets the effects of lactic acid, inducing the body to make adaptations to delay acidosis (increased acidity in the blood) and enabling the athlete to push harder without experiencing the same level of fatigue.

Research has shown that breath-holding exercises can improve an individual's tolerance to hypoxemia (low levels of oxygen in the blood) and reduce the acidity of the blood, eliminating oxidative stress and reducing lactic acid buildup. Athletes with long-term experience of breath-hold training, such as divers, have demonstrated a marked reduction in blood acidosis and oxidative stress in studies, suggesting that the extended practice of breath-hold exercises can be instrumental in avoiding the negative effects of free radicals produced by exercise.

Research spanning thirty years has investigated the mitigating factors in exercise-induced oxidative stress, taking into consideration different types of activity, duration, intensity, and the capabilities of the individual. The correct dose of physical exercise will of course vary from person to person according to physical condition and training habits, but the results of numerous studies show that oxidative stress can best be avoided by engaging in regular exercise combined with breath-hold training. The body is very good at adapting to consistent physical activity but cannot always react quickly enough to protect itself from a sudden influx of free radicals produced by infrequent high-intensity exercise. Exercising several times a week at a moderate, comfortable intensity from which you can recover easily is the best way to increase your body's natural antioxidant defenses and reduce oxidative stress. However, if you are a weekend warrior who does very little or no exercise during the week but engages in intense training on the weekend, you may be doing more harm than good.

More rigorous training regimens can also provide adequate protection from oxidative stress, so long as intensity and duration is increased gradually; a competitive athlete preparing for an upcoming event will need to allow a sufficient period of time to condition his or her body against oxidative stress. Studies show that well-trained athletes are perfectly able to cope with the oxidative stress caused by intense training and competition after the right kind of preparation—in fact, small amounts of oxidative stress can even prove beneficial for the body in strengthening antioxidant defenses.

While breathing naturally increases during physical exercise, individuals with a low BOLT score breathe more heavily than normal, creating an even greater quantity of free radicals relative to their physical work rate. A higher BOLT score, on the other hand, corresponds to a lower breathing volume, producing fewer free radicals and reducing the risk of muscle damage, injury, fatigue, and premature aging—possibly even extending the length of life. Breath-holding techniques offer a simple and effective way to increase antioxidant protection in athletes who exercise intensely and can easily be incorporated into regular training.

Alan is an amateur cyclist in his early twenties living on the west coast of Ireland. Very competitive by nature, Alan has won many races, often beating riders more seasoned and experienced than himself. He came to me for help because he found that it sometimes took him a half hour to recover his breathing after a race. Taking so long to catch his breath—even after strenuous activity—was a clear indication that Alan was pushing his body too hard during competition. As I expected, his BOLT score measured 15 seconds, meaning that his breathing volume was much larger than necessary. The breathlessness he experienced after exercise showed that his body was struggling to compensate. I explained to Alan that although he was clearly fit and capable of winning races, he was literally subjecting his body to abuse. At the time, he was suffering from a dry cough and head cold as a result of his exercise, but if he continued in the same way, the effects of overbreathing might not always be so mild.

My advice to Alan was to adapt his cycling to the capabilities of his body. First, he needed to increase his BOLT score to at least 35 seconds to align his breathing volume with his metabolic needs. I asked Alan to switch to nasal breathing during as much of his training as possible, reverting to mouth breathing only when absolutely necessary. Since the nose offers a smaller entry for breathing than the mouth, nasal breathing sets a limit to the volume of air taken into the lungs. Nasal breathing is a great barometer for training intensity, and my objective was for Alan to match his exercise to his capabilities by not pushing himself past the point where he could maintain nasal breathing. This is a safe and easy-to-implement approach, enabling a gradual but steady increase of BOLT score, which in turn allows for a sensible increase in intensity and duration.

One small land-based mammal has managed to confound the overwhelming evidence to support the negative effects of oxidative stress. For the past few decades, scientists have studied the naked mole rat—a bald, blind creature that looks like a hot dog with teeth and lives for up to twenty-eight years, almost eight times longer than any other rodent. The naked mole rat lives in East Africa, where it is considered a pest by local farmers as it burrows tunnels underneath fields and eats vegetable crops.

The breathing rate of the naked mole rat is very low in comparison to other rodents, and it lives in crowded colonies where there is little oxygen and high levels of carbon dioxide. As such, the naked mole rat is an excellent embodiment of the "less is more" theory of breathing. This might also explain how, despite living with high oxidative stress from a young age, the naked mole rat maintains good health and longevity, and in all the years this rather ugly animal has been studied, it has never been known to develop cancer. Even when scientists have injected the mole rat with cancer-causing agents, the disease was resisted. Exactly why the naked mole rat is immune to cancer is unknown, but some scientists are hopeful that finding an answer may provide the key to unlocking a cure for humans. Living congruently with nature's system of checks and balances seems to be the key to a long and healthy life. The naked mole rat does this remarkably well, given that researchers have discovered that the negative affects of high oxidative stress may be offset by high carbon dioxide levels.

Maintaining Your Fitness During Injury or Rest

The cost of an injury to an athlete can be devastating. Not only does the athlete have to endure the pain of the injury, but he or she also faces a drop in morale and the risk of decreased physical performance due to a lapse in training. Although a few days' rest from a regular exercise routine can lead to improved performance, several studies have shown that a rest period of around four weeks results in detraining effects on the body, including:

- Increase in body weight
- Increase in fat mass
- Increase in waist circumference
- Decrease of VO_2 peak

When you have worked so hard to increase VO_2 max and maintain your fitness, the effects of detraining can be enormously disappointing, especially when it becomes a recurrent event. For some, high-intensity exercise may be the source of a repetitive cycle of injury and detraining; when the body responds to injury with inflammation, free radicals are produced, which may lead to further muscle damage. There is, however, a way to prevent lapses in your regular routine, as well as maintaining fitness if you suffer an injury. The Oxygen Advantage program provides a solution to both the risk of injury and the limits an existing injury may place on your abilities. Practicing the Breathe Light to Breathe Right exercise along with breath holds can help to increase VO_2 max and the oxygen-carrying capacity of the blood while reducing lactic acid and improving blood flow. This optimal combination means that partial fitness can be maintained even in the case of injury or extended periods of rest.

A significant bonus of the Oxygen Advantage program is that it can be performed during rest or during exercise, and does not require an athlete to be injury-free. Some of the benefits of high-intensity exercise can even be obtained by adding breath-hold exercises to a gentle walk. Improving the way you breathe during rest and exercise will have positive repercussions on your general health as well as your athletic performance, reducing the risk of injury and enabling you to perform beyond your previous limits.

CHAPTER 11

Improve Oxygenation of Your Heart

On the morning of September 11, 2001, I received a phone call from my wife, Sinead, telling me to switch on the news. Listening to the accounts of what had taken place in New York City and at the Pentagon, I felt a chill go right through me. The tragedy felt even closer to home as Sinead and I had visited that wonderful city just three months prior.

On that same day, another tragedy unfolded, although it did not receive the same coverage as the terrorist attacks. In the United States alone, 3,000 people lost their lives to heart attack and stroke—two of the top three killers in the United States. The same tragedy occurred on Wednesday, September 12, and again on Thursday, September 13, repeating itself each and every day ever since. And while the falling of the Twin Towers will justifiably be remembered forever, the victims of cardiovascular disease are only remembered by their nearest and dearest. We cannot predict when a catastrophic event like 9/11 will occur, but we can help ourselves to extend and enrich the time we have to live and enjoy the company of those around us by taking care of our bodies and especially our hearts.

Understanding a simple and scientifically proven way of keeping our blood vessels healthy is invaluable to living life to the full. In this chapter, we explore the role of the gas nitric oxide, along with techniques for optimal breathing in order to maintain good cardiovascular health.

In 1867 the Swedish chemist, inventor, and industrialist Alfred Nobel invented dynamite by combining the chemical nitroglycerine with silica to form a less volatile explosive than nitroglycerine alone. Although his invention was initially intended to be used for blasting rock for industry, it later became synonymous with war and destruction. A few years after Nobel's invention, doctors discovered that this same chemical was effective in helping to reduce high blood pressure and treating the cardiovascular condition known as angina pectoris. In the human body, nitroglycerine—the same material used to make explosives—converts to the gas nitric oxide to provide amazing benefits for cardiovasular health. In his later years, Nobel suffered from heart disease, and when doctors tried to prescribe nitroglycerine to relieve his condition, he declined it, writing to his friend: "Isn't it the irony of fate that I have been prescribed nitroglycerine, to be taken internally! They call it Trinitrin, so as not to scare the chemist and the public." It was unfortunate that Nobel could not envision how a chemical so destructive outside the body could actually help it internally.

In 1896 Alfred Nobel suffered a stroke and died. In his will, the majority of his wealth was endowed to provide "prizes to those who, during the preceding year, shall have conferred the greatest benefit to mankind."

Although the motives of Nobel remain unclear, many commentators, including Albert Einstein, were of the view that this final act was an attempt to relieve his conscience and promote world peace. To ameliorate the negative effects from the invention of dynamite, Nobel ensured that a prestigious ceremony would take place each year to recognize those who make the greatest positive contributions to life.

In an ironic twist of fate, nearly one hundred years after Nobel's death, three doctors, Robert Furchgott, Louis Ignarro, and Ferid Murad, were awarded the Nobel Prize in Physiology or Medicine for discovering how nitric oxide had many important beneficial effects for the cardiovascular system. Had Alfred Nobel conceded to the wishes of his doctors, it is possible that his life may have been extended.

Sometimes referred to as the mighty molecule, nitric oxide is produced within the 100,000 miles of blood vessels throughout the human body, including the paranasal sinuses surrounding the nasal cavity.

Nitric oxide sends a signal for the blood vessels to relax and dilate. If there is too little nitric oxide, blood vessels constrict and the heart has to raise the pressure to send blood throughout the body. The easiest way to understand this is to imagine a garden hose with a knot in it: The water cannot flow freely and the pressure must rise if the water has any chance of flowing from one end to the other. Persistent high blood pressure or hypertension damages the arterial blood vessels, causing a buildup of plaque and cholesterol and also possibly blood clotting. If the blood clots and leads to an obstruction, this may cause the heart or brain to be deprived of blood and oxygen, resulting in a heart attack or stroke.

Nitric oxide plays a monumental role in human health by reducing cholesterol, reversing the buildup of plaque in the blood vessels, and helping to prevent blood clotting, all of which significantly increase the risk for heart attack and stroke. According to Nobel laureate and distinguished professor of pharmacology Dr. Louis Ignarro: "[Nitric oxide] is the body's natural defense to prevent all of these things from happening."

Producing sufficient nitric oxide enables blood flow to be directed effortlessly around the body, ensuring that vital organs receive sufficient oxygenation and nutrients. As blood vessels relax, the heart is able to normalize the pressure required to distribute blood throughout the body. Ways to increase nitric oxide include slow nasal breathing, regular moderate physical exercise, and eating foods that produce nitric oxide.

As nitric oxide is produced inside the paranasal sinuses as well as the blood vessels, breathing gently and calmly through the nose allows the gas to be picked up and carried to the lungs and blood. According to Jon Lundberg, professor of nitric oxide pharmacologics at the world-famous Karolinska Institute in Stockholm, Sweden, large amounts of nitric oxide are constantly released in the nasal airways of humans. As we breathe in through the nose, nitric oxide will follow the airflow to the lungs, where it plays a role in increasing the amount of oxygen uptake in the blood.

Dr. David Anderson of the National Institutes of Health in the United States also believes that how we breathe may well hold the key for how the body regulates blood pressure. It is well known that slow, gentle breathing from the diaphragm relaxes and dilates blood vessels, but the reasons behind this lasting drop in blood pressure is not completely understood. A plausible explanation is that the regular practice of relaxed breathing activates the body's relaxation response, resulting in improved blood gas regulation and dilation of blood vessels.

As we partake in physical exercise, blood flow increases and stimulates the inner lining of the blood vessels to produce more nitric oxide. An interesting study by a team of researchers from Hiroshima University Graduate School of Biomedical Sciences compared changes to blood flow in response to different intensities of physical exercise. Exercise intensity describes the perceived effort by individuals while physically moving their bodies. For example, most people will agree that walking at a moderate pace is a low-intensity exercise because it is easy to sustain and involves light demands in terms of breathlessness and recovery. The study, which was published in the journal *Circulation,* found that low-intensity exercise—which expends about the same amount of energy as window-shopping—wasn't enough to optimally increase blood flow. Conversely, high-intensity exercise—which includes vigorous activity at a fast pace—actually worsened blood flow. But the middle path—moderate-intensity exercise, such as a brisk walk or a light jog or cycle—increased production of nitric oxide and improved blood flow throughout the body.

While physical exercise is an excellent way to increase nitric oxide, diet, dietary supplements, and nasal breathing also play significant roles. In a recent conversation with Irish cross-country running coach John Downes, he told me how he actively encourages his athletes to drink beet juice, explaining how he witnessed an increase in performance and reduction of cramping as a result. Since John isn't a man to waste energy on unfruitful training practices, I decided to find out more. I soon discovered a study conducted by the University of Exeter that investigated the effects of increased dietary intake of beet juice, which is rich in the nitrates required to generate nitric oxide. A study group of men aged between nineteen and thirty-eight drank about two cups of beet juice every day for a week. This resulted in a "remarkable reduction" in the amount of oxygen required to perform exercise in comparison with a control group who drank water: The beet juice drinkers were able to cycle up to 16 percent longer before tiring. Furthermore, blood pressure within the beet juice drinkers dropped (within normal levels), even though it wasn't high to begin with. In conclusion, the researchers commented that the reduction of oxygen required for submaximal exercise following the drinking of beet juice "cannot be achieved by any other known means, including long-term endurance exercise training."

Along with beet juice, essential nitric oxide–producing, heart-protecting food sources to include in your diet include fish, green vegetables, dark chocolate, red wine (a glass per day—not the bottle!), pomegranate juice, green or black tea, and oatmeal. Food sources to be limited in your diet include the usual culprits of meat and processed foods. Along with eating the right foods, supplementing your diet with the amino acid L-arginine has been proven to increase nitric oxide production, although results vary depending on age and genetics. These simple changes to your diet, in addition to simply breathing lightly through your nose, may provide the key to lifelong cardiovascular health.

Most of us never give a moment's thought to our cardiovascular health, taking for granted that our heart will continue to perform its essential task for seventy years or more. But heart-related problems are not limited to those with a history of heart disease; completely avoidable cardiac issues can be experienced by young and otherwise healthy individuals, and prevented simply by increasing nitric oxide levels and by changing the way they breathe.

In 1909, American physiologist Dr. Yandell Henderson produced groundbreaking work on the relationship between breathing and heart rate that remains relevant today. In a paper entitled "Acapnia and

Shock: Carbon Dioxide as a Factor in the Regulation of Heart Rate," Henderson describes how he was able to regulate the heart rate of dogs to any rate he desired, from 40 beats or fewer per minute up to 200 or more, by altering their pulmonary ventilation. Henderson noted that even a "slight reduction of carbon dioxide of the arterial blood caused a quickening of the heart rate."

A few years ago, I worked with a thirtysomething woman named Anna who had been experiencing heart palpitations characterized by a fast heartbeat. Her resting pulse was about 90 beats per minute—the average heart rate is between 60 and 80 bpm—leaving her feeling that her "heart would beat right out of [her] chest." This feeling was a source of major distress for Anna and she had consulted a number of specialists, but there didn't seem to be anything physically wrong with her.

In an effort to get to the bottom of the problem, Anna had undergone a number of physical examinations and an electrocardiogram. The good news was that her cardiovascular health was normal. The bad news was that there was still no identified reason or solution for her predicament. Following her diagnosis, she was convinced that she had a medical condition incurable by modern science.

Unfortunately, Anna is not alone. The late chest physician Dr. Claude Lum wrote a number of articles illustrating her experience to a tee, based on patients who presented the very same pattern of symptoms with no physical anomaly. The one thing all these cases had in common was a tendency for overbreathing, the seemingly innocuous habit that is found to be the "mystery" cause behind numerous complaints and conditions in every field of medical practice.

Anna and her husband, having exhausted the traditional routes for a solution, somehow came across my work and were relieved to hear that the effects of overbreathing could indeed lead to palpitations characterized by an abnormally rapid heartbeat. With nothing to lose, they enrolled in my course.

When Anna arrived at the clinic, she appeared to be the picture of health, in her early thirties, slim, and of a petite build. For a few moments I observed her breathing without her noticing. She seemed to be a nose breather, but the one thing that caught my eye was that she would sigh every few minutes, lifting her shoulders and taking in a large breath. I'd seen the effects of regular sighing many times over the years, often in individuals prone to anxiety, and just like mouth breathing, it is a habit that usually goes unnoticed. I explained to Anna that in order to resolve her heart symptoms, it was very important that she retrain herself to stop this regular sighing.

Even though a sigh is often involuntary, taking place before the individual is aware of it, we still have a measure of control to reduce and eliminate the pattern. I explained to Anna that she should hold her breath or swallow any time she felt a sigh coming on. If by chance she missed one, she should hold her breath for 10 seconds to compensate for the overbreathing. I also provided her with a relaxation exercise and taught her the Breathe Light to Breathe Right exercise, which she proceeded to practice diligently for 10 minutes, 6 times per day. In addition, Anna began to pay more attention to her breathing throughout the day, ensuring that it stayed calm and quiet at all times.

When husband and wife returned one week later, Anna explained how much calmer she felt, and brought the fantastic news that her pulse had reduced to a perfectly average 60 to 70 beats per minute. Anna's case was one of my first experiences of the effect of overbreathing on cardiovascular health, and one that I will never forget—a clear demonstration of how overbreathing can affect us in so many different and potentially serious ways.

To demonstrate the effects of breathing on heart rate, I often ask my students to locate their pulse and take six or seven big breaths quickly through their mouth—within seconds, they are able to feel their pulse getting quicker. I then ask the students to practice gentle, slow, relaxed breathing and notice how the pulse slows down. If breathing rate and volume can have such an immediate and significant effect on the heart, we need to ask what repercussions poor breathing habits might have on the long-term health of our hearts.

The heart performs the most important function of the body, and, like all muscles, it requires sufficient blood flow and oxygenation in order to work properly. As Henderson showed, breathing in excess of normal metabolic requirements causes a reduced concentration of carbon dioxide in the blood.

This state of hypocapnia (which Henderson called *acapnia*) can affect cardiac functioning by decreasing the circulation of blood in the blood vessels and reducing blood flow to the heart. Since low levels of carbon dioxide in the blood lead to a strengthening of the bond between the red blood cells and oxygen, the result is reduced delivery of oxygen to the heart. On the other hand, increasing carbon dioxide levels in the blood by reducing breathing volume toward normal will result in improved blood flow and increased available oxygen, providing the heart with a ready and reliable supply of oxygen.

Cardiac Arrest in Athletes: A Missing Link

Each year, fit and healthy young athletes die from sudden adult death syndrome or cardiac arrest. These deaths have a far-reaching effect not only on family, friends, and classmates but also on entire communities.

Cormac McAnallen played Gaelic football for his native County Tyrone, winning almost every honor in the game during his career. He was also a student at Queen's University Belfast and University College Dublin, and was named Queen's University graduate of the year in 2004.

On March 2, 2004, at just twenty-four years old, he died suddenly in his sleep due to an undetected heart condition. Tributes to Cormac came from all sectors of society, including Irish president Mary McAleese, who hailed him as "one of the greatest Gaelic footballers of his time."

While undertaking research for this book, my curiosity was roused as to why healthy athletes might experience cardiac arrest or exhibit electrocardiogram (ECG) abnormalities with no other apparent risk factors. After all, most athletes are in the prime of their life, eat a good diet, do not smoke, have normal cholesterol levels and normal blood pressure, and generally care for their health. Aside from genetic predisposition, which of course we have absolutely no control over, what other factors might increase the risk of cardiac arrest in athletes?

In search of the reasons behind unexplained heart failure in young athletes, several studies have explored ECG abnormalities in order to find a connection between the electrical system that controls the heart's rhythm and unexpected cardiac arrest.

When the heart beats abnormally—either too fast or too slow, or irregularly—this condition is termed *arrhythmia*. Cardiac arrest happens when the electrical signals that control the timing and rhythm of the heartbeat become completely chaotic. When this happens the heart is no longer able to effectively pump blood around the body, and unless the condition is treated promptly, death is inevitable.

For the best chances of survival, immediate application of cardiopulmonary resuscitation (CPR) followed by defibrillation is essential. Although cardiac arrest often comes with no warning, clues such as abnormal heart rate, chest pain, dizziness, fainting, blackouts, and flu-like symptoms are sometimes present. Just prior to the onset of cardiac arrest, the athlete may feel dizzy or unwell and then collapse, stop breathing, and quickly lose consciousness as blood and oxygen stop flowing to the brain. Unless circulation is restored within a few minutes, irreversible brain damage and subsequently death will result from sudden heart failure and the cessation of circulation.

An electrocardiogram (ECG) is a test used to interpret the electrical activity of the heart, assessing the rate and regularity of the heartbeat as well as the presence of any damage to the heart muscle. In assessing ECG abnormalities, doctors examine various indicators that are linked to a number of life-threatening cardiac conditions.

Studies have found that certain ECG changes in young athletes are common and usually reflect adaptations of the heart as a response to regular physical training. However, certain abnormal ECG readings, such as T-wave inversion and ST segment depression, were found to be potential precursors to sudden and unexpected cardiac arrest during sport or exercise. Markedly abnormal ECGs in young and apparently healthy athletes may also suggest the initial signs of underlying heart disease.

ST segment depression is considered to be a sign of reduced circulation in the blood vessels of the heart, and it has been suggested that there is a link between ST segment depression and the risk of sudden cardiac death. In a study including 1,769 men without evident coronary heart disease, a total of 72 deaths occurred in the eighteen years of follow-up—all of whom showed asymptomatic ST segment depression during exercise on their ECG readings.

Earlier on we discussed how overbreathing reduces blood flow and oxygen delivery to the heart. An opportune question at this point is whether the amount of air we breathe plays a role in the onset of cardiac arrest. This, I think, could be an important factor in the investigation of sudden cardiac death in young athletes.

A study conducted by researchers from the University of Patras in Greece revealed how the amount of air we breathe can produce changes in electrocardiogram results. During the study a total of 474 healthy volunteers with no apparent heart disease increased their breathing rate to more than 30 breaths per minute for 5 minutes to create the effects of hyperventilation. ECG readings reported abnormalities in 72 volunteers, including findings of ST depression and T-wave inversion, with 80.5 percent of abnormalities occurring within the first minute of hyperventilation. Interestingly, the study found that age, gender, smoking, and hypertension did not influence the overall incidence of the abnormalities, showing that even perfectly healthy individuals can be susceptible to the abnormalities caused by hyperventilation.

If increasing the breathing rate to 30 breaths per minute over 5 minutes can induce ECG abnormalities, what effects might strenuous exercise have on an athlete's risks of cardiac disease when you consider that air intake can increase to between 50 and 70 breaths per minute during moderate- to high-intensity activity? Should athletes be taught how to ensure healthy breathing volume during exercise in order to minimize the effects of hyperventilation on their cardiovascular health?

Penny is a cardiac nurse who has worked at a Limerick hospital for the past thirty years. As a fit and healthy woman of normal build, Penny became worried when she began to experience symptoms of arrhythmia in her sixties. The issue had developed gradually over a few years, and Penny described the feeling as "a large butterfly fluttering about in the left side of my chest," which could occur at any time of the day or night, sometimes going on for eight hours or more.

It all started when Penny's job became more demanding and she took on extra responsibilities and hours. Ireland has experienced an economic crisis for several years, leading to many cutbacks in our health service. As a result, the frontline staff of nurses has taken much of the brunt in the form of extra workloads. For Penny, the effect of these changes was increased anxiety, which she saw as the main cause of her arrhythmia.

As an episode of arrhythmia began, Penny would feel a need for extra oxygen. To satisfy this air hunger, her anxiety and breathing would increase, making her heart race and further contributing to her complaint. It was a vicious circle, as her symptoms fed back into the condition.

I met Penny at my clinic in Limerick and observed that she breathed through both her nose and mouth. Her breathing was noticeable, from the upper chest, and there was no natural pause on the exhalation. Her BOLT score was 8 seconds, leaving me in no doubt that she was chronically hyperventilating: the possible source of her cardiac problems.

To begin retraining Penny's breathing, I helped her to learn how to breathe using her diaphragm. I instructed her to place one hand on her chest and one hand just above her navel so that she could easily feel where her breathing was coming from, and begin to direct her breathing into her abdomen. Breathe in: abdomen out. Breathe out: abdomen in. Penny's next step was to exert a small amount of pressure against her chest and abdomen with her hands so that she felt a slight resistance to her breathing. In blocks of 3 minutes, Penny practiced calming her breathing, gently slowing it down and taking less air into her body in order to create a slight shortage of air. I asked Penny to practice this exercise for 10 minutes, 5 times per day. The rest of her simple program consisted of breathing through her nose at all times and wearing paper tape across her mouth at night to make sure she didn't breathe through her mouth during sleep.

I met Penny several times over the following weeks, and by the third week her BOLT score had increased to 25 seconds. More important, her symptoms of arrhythmia had reduced significantly.

The exercise I gave Penny is very similar to the Papworth method developed by Dr. Claude Lum. Dr. Lum was well known for his studies of overbreathing and was described as an "archetypal caring physician" who displayed the rare qualities of both sympathy and patience, particularly to those with psychosomatic disease. In 1959, at Papworth Hospital in Cambridgeshire, England, Dr. Lum formed part of a team developing cardiopulmonary bypass techniques, and during the following decades, his interest in habitual hyperventilation burgeoned. Together with his team of physiotherapists, he developed the Papworth method in order to address this common breathing disorder. Dr. Lum dedicated every effort to generate greater awareness of hyperventilation syndrome through his writings and lectures, many of which were published in reputable medical journals, including the *Lancet,* the *Journal of the Royal Society of Medicine,* and the *Journal of Psychosomatic Research.* He was one of those rare doctors with the drive and courage to devote much of his working life to unearthing the cause of so many common diseases of civilization that are, at best, only just managed with medication.

Heart Attack: A Missing Link

Myocardial infarction, otherwise known as a heart attack, occurs when blood flow to the heart is severely reduced or cut off altogether. This stoppage of blood results in oxygen starvation and damage or death to part of the heart muscle.

Heart attacks often occur during or following physical exercise or emotional stress. Both activities increase breathing volume, and when breathing volume is greater than the body's metabolic needs, carbon dioxide is removed from the lungs and blood, resulting in reduced blood flow and reduced oxygenation of the heart.

Up to 10 percent of heart attack patients have symptoms attributable to hyperventilation. In one particular study, 3 to 6 percent of patients showed normal findings on coronary angiography soon after they experienced myocardial infarction, suggesting that the infarctions were not in fact due to any underlying heart disease but could have resulted from hyperventilation.

Reduced blood flow to the heart muscle due to hyperventilation may then be partially or wholly responsible for myocardial infarction in some individuals. It follows, therefore, that the way we breathe, and the resultant levels of carbon dioxide in our blood, can have significant effects on the health and function of our hearts.

In the following sections we will investigate whether patients with cardiac problems, including those who suffered a heart attack, breathe more heavily than normal and whether breathing exercises aimed at correcting breathing volume may reduce the risk of further cardiac problems, and whether hyperventilation during resuscitation can adversely affect outcomes.

Heart Disease and Hyperventilation

People with some types of heart disease tend to breathe more heavily and more intensely than healthier people, but many also experience a reduction of symptoms when their breathing volume is corrected toward normal. If these individuals had practiced light breathing in the first instance, would they have been less at risk of developing heart disease?

A study of twenty patients with moderate to severe chronic cardiac failure showed that these individuals had a breathing volume of between 15.3 to 18.5 liters per minute. Given that normal breathing volume should be between 4 to 6 liters per minute, each of these patients was breathing a volume of air that was enough for two or three people. This research, along with other similar studies, show that patients with chronic heart failure breathe too intensely. Patients who exhibited heavy breathing were also found to have feelings of breathlessness during physical exercise. This comes as no surprise when you realize that how we breathe during rest determines how we breathe during physical exercise. Breathing noticeably from the upper chest during rest leads to increased breathlessness during physical exercise and the cycle of overbreathing is destined to continue.

It is evident from this research that the way we breathe is a contributing factor in cardiac health, showing a positive correlation between increased breathing volume and the severity of chronic heart failure. Not only does overbreathing reduce the ability of the heart to pump blood around the body, an excessive breathing volume also reduces blood flow to part of the heart muscle, causing insufficient oxygenation. In a 2004 study published in the *European Journal of Cardiovascular Prevention and Rehabilitation,* fifty-five men were examined two months after suffering a heart attack. After following a program of breathing exercises, the patients' breathing volume per minute significantly decreased by approximately 50 percent—from 18.5 to 9.8 liters. Remembering that normal breathing volume per minute is 4 to 6 liters, it is evident from this study that patients who suffered from a heart attack also tend to breathe far in excess of what is required, but that this volume can be reduced much closer to normal simply by implementing correctional breathing exercises.

In addition, patients who practiced these breathing exercises showed an increase in the concentration of carbon dioxide in their arterial blood, from 33.2 mmHg to the highest point of the normal range of 44.2 mmHg. Based on improvements to breathing volume and respiratory function, the authors of the study recommended that breathing retraining could act as a valuable rehabilitation measure after heart attacks.

Other studies confirm these benefits, showing how breathing exercises can have lasting effects on respiratory function and help to reduce symptoms of cardiac dysfunction.

Hyperventilation During Cardiopulmonary Resuscitation (CPR)

We have seen clearly how the effects of breathing light can improve blood flow and oxygenation, and may even be able to help prevent heart attacks in those who breathe excess volumes of air. Overbreathing can cause numerous health concerns, but there is a more disturbing risk associated with this condition that could literally make the difference between life and death.

Cardiopulmonary resuscitation (CPR) is performed during cardiac arrest to help preserve normal brain function until further measures can be taken to restore blood circulation and breathing. We know that hyperventilation causes reduced blood flow and reduced oxygenation of the heart, but studies have also revealed that excessive ventilation during CPR is actually detrimental to survival.

Researchers investigated instances where CPR resulted in death due to excessive ventilation applied by well-trained but overzealous rescue personnel. Despite adequate training, these professionals hyperventilated their patients while attempting resuscitation through higher than necessary breathing rates. It is thought that the high airway pressure resulting from administering more air into the patient than necessary had a detrimental—and ultimately fatal—effect on patients' blood flow. One study concluded with the following warning: "Additional education of CPR providers is urgently needed to reduce these newly identified and deadly consequences of hyperventilation during CPR."

Reviewing the findings above, it is shocking to think that the very procedure designed to help save lives might in fact be having the opposite effect. It is even more shocking when we consider that the relationship between breathing volume and blood flow to the heart was first documented over a century ago. Thankfully, since 2007 there has been a monumental change in CPR procedures involving manual ventilation. Increasingly, the emphasis during CPR is now on chest compressions to maintain circulation, rather than manual ventilation.

Over the years, I have witnessed many young athletes at all levels of fitness breathing too intensely for their given level of exercise. In this chapter, I have attempted to join the dots between excessive breathing volume, reduced oxygenation of the heart, and resultant ECG abnormalities, heart attacks, and chronic heart disease. It is only logical to surmise that a poorly oxygenated heart is less able to cope with the demands of intense physical exercise. Yet every month I see reports of children, teenagers, and young adults who were in the prime of life dying from undetected heart conditions. Upon hearing the news I often wonder: Could this tragedy have been avoided if the victim had been encouraged to breathe normally and through the nose? The breathing volume of athletes and nonathletes alike is crying out for attention, and greater awareness would be worth all of the effort, if even just one young life were saved.

CHAPTER 12

Eliminate Exercise-Induced Asthma

In childhood, treatments for forty-three-year-old Julian's asthma included cough medicine, trips to the coast for the benefits of sea air, and inhaling steam from a boiling kettle. On some nights, Julian remembers his wheezing being so bad that he stayed up most of the night with his head out the window in an effort to breathe. Any individual who experienced asthma as a child during the 1970s and 1980s can probably testify to the lengths that similarly worried parents took to try to help their children to breathe.

By the late 1980s Julian had been prescribed various relievers and preventative medications, in addition to regular trips to the hospital for nebulizer treatment. This never-ending cycle of medication and hospitalizations continued for many years, and although Julian tried to keep fit, he often found he was unable to breathe, particularly during the small hours of the night.

Fast-forward to 2006, and Julian was taking higher doses of asthma medication while his fitness level gradually reduced, a totally unproductive cycle that was beginning to seriously affect his health and well-being. Julian's story is typical of any individual with moderate to severe asthma; although physical exercise can be very beneficial, many persons with asthma simply tend to avoid it for fear of having an attack.

In early 2007, Julian attended one of my Dublin courses at which we focused on breathing through the nose, breathing light, and practicing breath holds while walking. Julian took his last dose of reliever medication the day after the course.

Within six months, Julian's asthma had drastically improved, and by Christmas 2007, he had his last dose of preventative medication. His fitness also improved and he was able to swim a mile per day, five times a week. In 2008, Julian's GP agreed to reclassify his medical record as "asthma resolved."

Over the next three years, Julian's exercise plan evolved to include eight hours of high-intensity indoor cycling, circuits, and stretching classes per week, as well as the techniques of nasal and reduced-volume breathing he learned from my course.

These changes, along with adjustments to his eating habits, allowed Julian to improve his performance and enjoy more energy and stamina at a higher level of activity. At the age of forty in 2012, Julian ran five half-marathons and covered over 750 training miles. He achieved a personal best of 1:46 in his third half-marathon; two weeks later he completed the Berlin full marathon in 3:57. Following the Berlin Marathon, he ran the Dublin City Marathon in 4 hours. In six months, Julian had knocked over 8 minutes off his first half-marathon time.

In six years, Julian progressed from a chance encounter with one of my books to attending my breathing course, improving his fitness, completely eliminating prescription medicine for his asthma, and running half and full marathons in very respectable times!

The word *asthma* derives from Greek and means "to pant." While asthma has been around for a very long time, it affects more people today than ever before. Exercise-induced asthma affects an estimated 4 to 20 percent of the general population and 11 to 50 percent of certain athlete populations. Interestingly, one study showed that while 55 percent of football athletes and 50 percent of basketball players displayed airway narrowing conducive to asthma, athletes from the sport of water polo showed significantly fewer asthma symptoms. Later on in this section we will investigate why this might be.

So, what causes asthma? The most common theories include the hygiene hypothesis, which rests on the premise that too much cleanliness means children are not exposed to enough germs, resulting in diminished immune capabilities later on in life. A second commonly cited explanation is an increase in pollution, but while this may well be a trigger, it is not necessarily the cause. For instance, the west of Ireland, where I live, has a high asthma rate but very good air quality.

Might there be another factor that plays a significant role in causing asthma—that of habitually breathing too much? If this were true, then surely reducing breathing volume could result in a reversal of the condition. By looking at the causes and symptoms of asthma, and the physiological changes resulting from the condition, we can begin to determine just how important breathing exercises can be in treating asthma.

Since asthma is a condition characterized by difficulty in breathing, a logical approach would be to attempt to find the root cause by first addressing poor breathing habits. Tackling asthma from this angle is not new and was employed by the ancient Greek physician Galen and the sixteenth-century doctor Paracelsus, who recommended breath holding and breathing exercises for the treatment of coughing and narrowed airways.

The prevalence of asthma increases relative to wealth. Increased wealth leads to a change in living standards: Food becomes more processed, competitive stress increases, houses become more airtight, we perform less physical exercise, and the majority of our jobs are sedentary. Fifty years ago, our living and working situations were quite different, and asthma rates were significantly lower. During that time, we ate more natural foods, had less competitive stress, our houses were drafty, and most occupations involved physical labor. Back then, our lifestyle was conducive to a more normal breathing volume, and, as a result, asthma was far less common.

As we have seen, normal breathing volume for a healthy adult is generally agreed to be 4 to 6 liters of air per minute, but adults with asthma demonstrate a resting breathing volume of 10 to 15 liters per minute, two to three times more than required. Imagine the effect on the respiratory system when an individual breathes twice or three times too heavily all day, every day.

Normal breathing during rest involves regular, silent, abdominal breaths drawn in and out through the nose. People with asthma, on the other hand, display habitual mouth breathing with regular sighing, sniffing, and visible movements from the upper chest. During an exacerbation of asthma, symptoms like wheezing and breathlessness increase along with respiratory rate, relative to the severity of the condition. In other words, as asthma becomes more severe, there is also an increase in breathing volume.

While it is well documented that people with asthma breathe too much, there is a need to determine whether the increase to breathing volume is a cause or effect of the condition. As the airways narrow a feeling of suffocation is generated, and a normal reaction is to take more air into the lungs to try to eliminate this sensation. Either way, it is a vicious cycle; narrowed airways lead to heavier breathing that causes an increase in breathing volume, resulting in the narrowing of the airways and on and on, worsening the condition and establishing bad breathing habits as a matter of necessity.

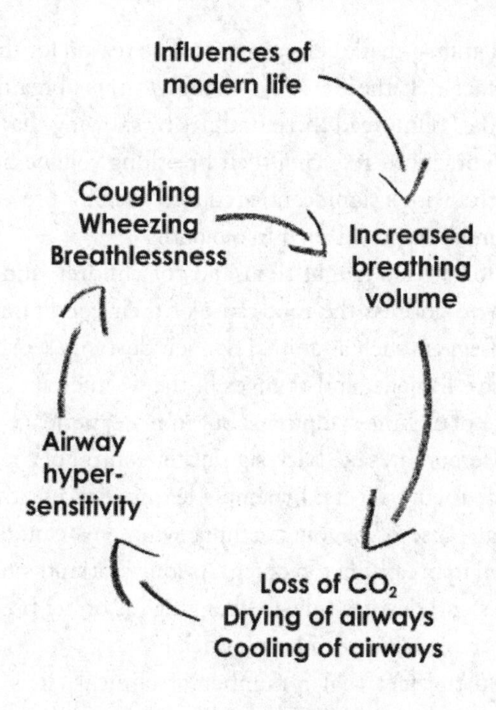

The only way to determine whether breathing too much causes asthma is to investigate what happens when a group of individuals with asthma practice breathing exercises designed to bring their breathing volume toward normal.

A study at the Mater Hospital in Brisbane found that when the breathing volume of adults with asthma decreased from 14 liters to 9.6 liters per minute, their symptoms reduced by 70 percent, the need for rescue medication decreased by 90 percent, and the need for preventer steroid medication decreased by 50 percent. The study found a direct relationship between the reduction of breathing volume and improvement to asthma. The closer breathing volume reduced toward normal, the greater was the reduction of asthma symptoms such as coughing, wheezing, chest tightness, and breathlessness. Furthermore, the trial's control group—who were taught the hospital's in-house asthma management program—made zero progress. The reason for this was solely due to the fact that there was no change to their breathing volume. Further studies reinforced these findings by showing that people with asthma who practiced reducing their breathing volume had far better asthma control with a significantly reduced need for preventive steroid and rescue medication within 3 to 6 months.

Since 2002 I have taught thousands of children and adults with asthma how to address the root cause of their condition—breathing too much. Triggers such as animal dander, dust mites, exercise, pollution, excessive hygiene, and changes in the weather are often cited to be the cause of asthma symptoms, but in my experience the vast majority of sufferers can take back significant control of their condition, regardless of their triggers, by simply learning to breathe light. The fundamental cause of ongoing asthma symptoms can almost always be attributed to breathing too much. As long as a student understands the exercises and devotes time to changing his or her breathing, positive results are consistent and reproducible.

Based on the fact that a number of clinical trials have shown that asthma symptoms and the need for asthma medication are significantly reduced following the employment of reduced breathing exercises, there is no doubt that overbreathing is a significant contributor to asthma. Of course, it is also normal for people prone to asthma to increase their breathing volume to compensate for a feeling of suffocation, but this action is simply part of a feedback loop. Elements of modern living increase breathing volume, which can activate asthma in genetically predisposed individuals. As their asthma takes hold, the individual breathes faster and more intensely, worsening the condition. While it is important to recognize this feedback loop, the first step to addressing asthma is to reduce excessive breathing habits.

I can relate to any child or adult with asthma because for more than twenty years I struggled with the same symptoms; I was unable to perform even the most basic physical exercise, my nose was constantly stuffed, and I continuously breathed through an open mouth. Year after year my asthma medication increased with no signs of my symptoms abating. My sleep, concentration, mood, and quality of life were all adversely affected. It was only by chance, when I learned of the work of Russian doctor Konstantin Buteyko, that I was able to reverse my asthma. Within a few short days my wheezing decreased dramatically, simply by learning to unblock my nose and normalize my breathing volume. I have now been wheeze-free for the past thirteen years, and all I did was to learn to breathe normally again.

This had such a tremendous impact on my life that I changed careers in 2001 and retrained under the auspices of the late Dr. Buteyko. In 2002, I founded Asthma Care to help publicize this information to children and adults with asthma. Our clinics are now offered in a number of countries.

The first step to addressing chronic overbreathing is to make the switch from mouth to nasal breathing. While nasal breathing is important for everyone, for people prone to asthma it is vital. When breathing volume is greater than normal, there is a tendency to open the mouth in order to allow more air to enter the lungs. People diagnosed with asthma often feel they are not taking in enough air while breathing through the nose, which causes them to breathe through the mouth.

Mouth breathing influences asthma in a number of ways:

- Air taken in through the mouth is not filtered of airborne particles, including germs and bacteria.
- The mouth is simply not as effective as the nose in conditioning air to the correct temperature and humidity prior to entering the lungs.
- Because the mouth provides a larger space to breathe through than the nose, breathing volume will be higher, causing too much carbon dioxide to be expelled from the lungs. Carbon dioxide is a natural "opener" of the

smooth muscle in the airways. The loss of carbon dioxide therefore causes asthma airways to narrow even more.

- Unlike nasal breathing, mouth breathing does not allow us to benefit from nasal nitric oxide, which supports the lung's defensive capabilities.

Taking all these factors into consideration, it is not surprising that mouth breathing causes a reduction in lung function in people with mild asthma and plays a significant role in the exacerbation of asthma symptoms.

Not only is it important to breathe through the nose during rest, it is also beneficial to nasal breathe during physical exercise. In a paper published in the *American Review of Respiratory Disease,* researchers studied the beneficial effects of nasal breathing on exercise-induced asthma. The study observed that most subjects with asthma spontaneously breathed with their mouths open when instructed to breathe "naturally." The authors found that mouth breathing during exercise caused the airways to narrow even further. In contrast, when subjects were asked to breathe only through their nose during exercise, exercise-induced asthma did not occur at all. The paper concluded that "the nasopharynx and the oropharynx play important roles in the phenomenon of exercise-induced bronchoconstriction." In simple terms, the effects of breathing through the nose are integral to reducing or avoiding exercise-induced asthma completely.

The fact that elite athletes with asthma often favor swimming above other forms of exercise is not a coincidence. During swimming, the face is immersed underwater, reducing the amount of air taken into the lungs and increasing the athlete's tolerance for carbon dioxide. Although the swimmer may draw his or her breath in through the mouth, the protective effects of reduced breathing are still evident. A child or adult with asthma may also prefer swimming because the water exerts a gentle pressure on the chest and abdomen, further restricting breathing volume and improving athletic performance.

The difference between land-based exercise and swimming in terms of breathing pattern and volume is significant for people with asthma. On land, your breathing pattern during exercise is not restricted the way it is in water, meaning that you can very easily overbreathe, resulting in constricted airways, a reduction in the amount of CO_2 in your blood, and a lower BOLT score. For an individual with asthma, overbreathing during rest leads to overbreathing during exercise, which in turn leads to exercise-induced asthma. However, exercising in water naturally causes you to restrict your breathing and lower your breathing volume toward normal, providing a much safer and more productive environment for people with asthma to exercise.

At the beginning of this chapter we looked at statistics that showed that in a group of athletes, narrowing of the airways affected 55 percent of football athletes and 50 percent of basketball players, but 0 percent of water polo players. With such a glaring disparity, what factors could possibly explain the difference? The answer, as you have probably guessed by now, is simple. Water polo training involves breath holding and swimming underwater, resulting in a higher tolerance to carbon dioxide, increased amounts of nitric oxide, and a reduced breathing volume. With a more normal breathing volume, asthma tendencies don't appear.

However, if you have asthma and do not wish to take up swimming, there is a simpler way! The Oxygen Advantage approach incorporates all of the beneficial aspects of swimming and more. Although the act of swimming has its merits, it is well documented that spending time in chlorinated pools is not ideal for asthma, as the chlorine can cause damage to lung tissue. Furthermore, while swimming reduces breathing volume, it is still important to address poor breathing habits outside of the pool. Many swimmers remain habitual mouth breathers and continue to employ poor breathing habits that reduce their athletic performance and maintain their asthma.

Your success in addressing asthma will be based on your ability to increase your BOLT score using the Breathe Light to Breathe Right and simulation of high-altitude exercises described in this book. You will find a program specific to your needs in Part IV. The general aim is to increase your BOLT score to 40 seconds. The best time to measure your BOLT score to track your progress is first thing in the morning, and if your BOLT measurement remains less than 20 seconds, then your asthma symptoms will persist. However, when your early morning BOLT score is greater than 20 seconds, symptoms such as wheezing, coughing, breathlessness, and chest tightness will disappear. It's important to note that you may still be susceptible to certain symptoms even when you have achieved a BOLT score of 20 seconds when exposed to a trigger; a BOLT score of 40 seconds is needed to fully eliminate your asthma symptoms.

As you work toward achieving a high BOLT score, you may continue to experience symptoms, depending on your medical history and triggers. Your ability to stop the effects of asthma using the exercise below is dependent on two factors: how high your resting BOLT score is and how quickly you react to the onset of symptoms. The sooner you begin practicing the exercise, the easier it is to prevent symptoms from taking hold. By ignoring your symptoms and hoping they will go away by themselves, the effects of asthma tend to get worse and can take on a life of their own. If you often experience asthma symptoms, you will know that wheezing and coughing usually worsen over time, so it is important to intervene early on.

This exercise can help stop asthma symptoms, but please get your doctor's permission to try it first. Then follow the instructions below during the early stages of chest tightness, wheezing, coughing, or a head cold. If you aren't able to stop your symptoms within 10 minutes, then take your rescue medication. If you are experiencing severe symptoms, of course, then take rescue medication straightaway. If your rescue medication does not stop your symptoms within a couple of minutes, it is advisable to call a medical doctor immediately.

To stop asthma symptoms before they take hold, follow these steps:

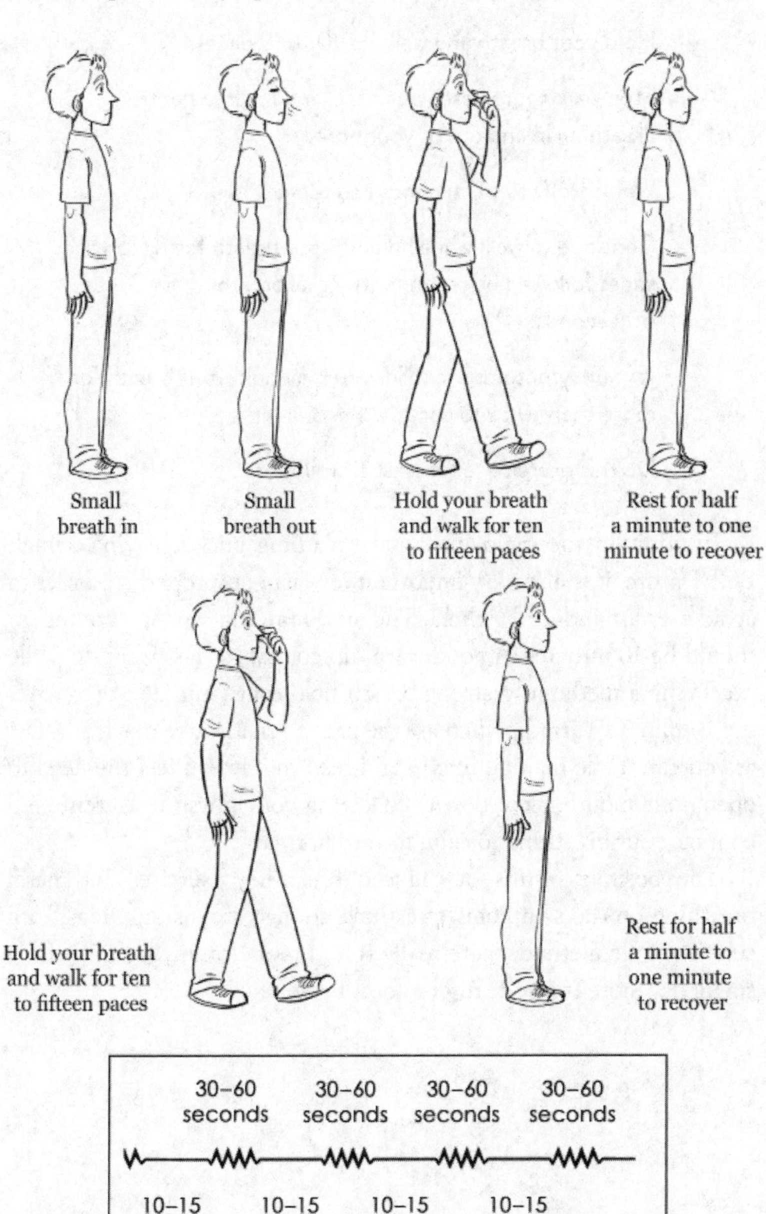

- Take a small, silent breath in and out through your nose.
- Hold your breath and walk for 10 to 15 paces.
- Stop walking, release your nose, and resume gentle breathing in and out of your nose.
- Wait for 30 to 60 seconds and repeat.
- Continue to walk while holding your breath for 10 to 15 paces followed by resting with nasal breathing for 30 to 60 seconds.
- If your symptoms are mild, you may hold your breath for more than 10 to 15 paces.
- Do this exercise for at least 10 minutes.

In addition to employing nasal breathing and achieving a high BOLT score, it is also very important to warm up properly in order to avoid exercise-induced asthma. The minimum time spent warming up should be 10 minutes. A good warm-up consists of fast walking while practicing a medium to strong breath hold every minute or so. After the 10-minute warm-up, increase the pace so that you're moving as fast as you can while maintaining nasal breathing. If you feel the need to open your mouth, slow down. Following your physical exercise, encourage your breathing to calm toward normal.

The positive results of reduced breathing exercises and nasal breathing on the symptoms of asthma are fast and astounding. With such a simple method, there really is no reason for anyone to suffer a single day more from the discomfort of asthma.

CHAPTER 13
Athletic Endeavor—Nature or Nurture?

In 1704, a racing stallion by the name of the Darley Arabian arrived in Britain from Syria, and 95 percent of today's male thoroughbreds descend from him. Geneticist Patrick Cunningham and colleagues from my alma mater, Trinity College Dublin, traced the lineage of nearly one million horses from the past two centuries and determined that 30 percent of variation in performance in thoroughbreds is due to genetics alone. In the nature versus nurture debate, these results suggest that nature plays a significant part in our athletic abilities.

There is one area in particular where a combination of genetics and behavior has considerable influence on athletic performance, and that is the way the face and jaws develop during childhood. For example, take a look at the structure of the face and jaws of past Olympic gold medal winners including Usain Bolt, Sanya Richards-Ross, Steve Hooker, and Roger Federer. What is strikingly apparent for this group, and for the vast majority of top-class athletes, is the forward growth of the face and width of the jaws. Athletic success depends on having good airways, which in turn is dependent on normal facial structure. Spend a lot of time with your mouth hanging open or sucking your thumb during childhood and the face grows differently from how nature intended.

In fact, Michael Phelps, the most decorated Olympian of all time, is one of the very few top-class athletes who does not exhibit forward growth of the jaws and a wide facial structure. Based on his facial profile, there is a high likelihood that he was a mouth breather during childhood, possibly requiring orthodontic treatment in his early teens. It is also possible that Phelps chose swimming, either consciously or unconsciously, as it was the one sport that he could excel in. The very act of swimming restricts breathing to help offset any negative effects that have developed from mouth breathing or an inefficient breathing pattern.

Although the natural order of things is to breathe through the nose, many children—especially those with asthma or nasal congestion—habitually breathe through the mouth. Brazilian researchers investigating the prevalence of mouth breathing in children aged 3 to 9 found that 55 percent of a random selection of 370 subjects were mouth breathers. Children who regularly breathe through their mouth tend to develop negative alterations to their face, jaws, and the alignment of their teeth. Mouth breathing affects the shape of the face in two ways. First, there is a tendency for the face to grow long and narrow. Secondly, the jaws do not fully develop and are set back from their ideal position, thus reducing airway size. If the jaws are not positioned forward enough on the face, they will encroach on the airways. See for yourself: Close your mouth, jut out your chin, and take a breath in and out through your nose, noting the way air travels down behind the jaws. Now do the same but pull your chin inward as far as you can—you will probably feel as if your throat is closed up as you try to breathe. This is exactly the effect poorly developed facial structure has on your airway size. It is no wonder that those with restricted airways tend to favor mouth breathing.

The forces exerted by the lips and the tongue primarily influence the growth of a child's face. The lips and cheeks exert an inward pressure on the face, with the tongue providing a counteracting force. When the mouth is closed, the tongue rests against the roof of the mouth, exerting light forces that shape the top jaw. Because the tongue is wide and U-shaped, it follows that the shape of the top jaw should be wide and U-shaped also. In other words, the shape of the top jaw reflects the shape of the tongue. A wide U-shaped top jaw is optimal for housing all our teeth.

However, during mouth breathing, it is very unlikely that the tongue will rest in the roof of the mouth. Try it for yourself: Open your mouth and place your tongue on your upper palate. Now try to breathe through your mouth. While it is possible to draw a wisp of air into the lungs, it will not feel right. It follows therefore that the tongue of a mouth breather will tend to rest on the floor of the mouth or suspended midway. Since the top jaw is not then shaped by the normal pressures of the tongue, the end result is the development of a narrow V-shaped top jaw. Aesthetically, this contributes to a narrowing of the facial structure, crooked teeth, and orthodontic problems. It has been well documented that mouth-breathing children grow longer faces.

The second way facial structure is affected by the way we breathe during childhood is the position of the jaws. The way the jaws develop has a direct influence on the width of the upper airways. Our upper airways comprise the nose, nasal cavity, sinuses, and throat. Strong athletic performance requires large upper airways that enable air to flow freely to and from the lungs. Although a high BOLT score and effective breathing technique are crucial for high levels of performance, having airways that function with little resistance is also very advantageous. For example, a marathon runner who has efficient breathing but airways that are the width of a narrow straw is not going to get too far.

The normal growth of the face is forward. Since a mouth-breathing child does not rest his or her tongue in the roof of the mouth, the jaws are unable to be properly shaped by the tongue, and the natural forward growth of the jaws is impeded. This results in jaws that are set back from their ideal position, compromising airflow. For the correct development of the lower half of the face and airways, it is imperative that a child habitually breathes through his or her nose. Breathing through the nose with the tongue resting in the roof of the mouth helps to establish the ideal conditions for the normal development of the face.

I switched from mouth to nose breathing during the late 1990s, when I was in my early twenties, but it was only after I met with myofunctional therapists Joy Moeller, Barbara Greene, and Karen Samuel in 2006 that I learned the correct position of the tongue. Until then, I hadn't given it a moment's consideration, and most likely my tongue had been flopping about without a home for the previous thirty-two years. Between them, Joy, Barbara, and Karen have devoted almost a hundred years to reeducating people on the placement of the tongue and facial muscles in order to address a variety of detrimental issues that affect the development of the jaws and teeth. Spending thousands of dollars on orthodontic treatment can be in vain if poor habits such as mouth breathing, tongue thrusting, and incorrect swallowing are not addressed. And you may be able to avoid orthodontic treatment altogether if these habits are not permitted to develop in the first place.

In the correct resting position, three-quarters of the tongue should press gently against the roof of the mouth with the tip of the tongue placed just behind the top front teeth—the same place we put the tongue to make the *N* sound *"nuh."* Just like nasal breathing, optimal resting tongue posture is not a recent discovery; for thousands of years it formed an important part of Eastern yoga and the religion of Buddhism. Yogi Bhajan, who introduced Kundalini yoga to the United States in 1968, accredited the upper palate and the tip of the tongue as the two most important parts of the body. The ancient Buddhist scriptures of the Pali Canon contain passages describing how the Buddha pressed his tongue against the roof of the mouth for the purposes of controlling hunger and the mind.

This illustration shows the facial characteristics of a nasal breather and is based on the Irish International and LA Galaxy soccer captain Robbie Keane:

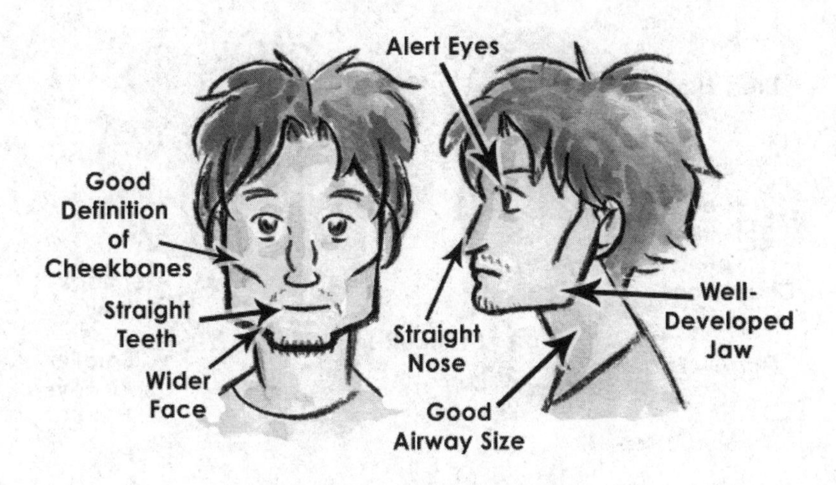

Note the forward position of the jaws, high cheekbones, airway size, and width of the face. The jaw is strong and positioned well forward so that the chin is nearly as far forward as the tip of the nose. When cartoonists draw illustrations of a dominant male, his strength is often conveyed by a rugged and exaggerated jaw. Socially, a wide facial structure and strong jawline are considered healthier and more attractive than a recessed chin. And not only is the classic square jawline more likely to get you a date, it might also be beneficial to your bank balance. In a paper written by researchers from the School of Business at the University of California at Riverside, men with wider faces were found to be stronger negotiators, commanding a signing bonus of nearly $2,200 more than their narrow-faced counterparts. In a separate study by the same authors it was found that companies led by men with wider faces also achieved superior financial performance.

Throughout our evolution, social anthropologists have regarded facial appearance as a determinant in establishing societal rank and individual roles. Beauty is not just skin deep, and Aristotle was right when he stated that "Beauty is a far greater recommendation than any letter of introduction."

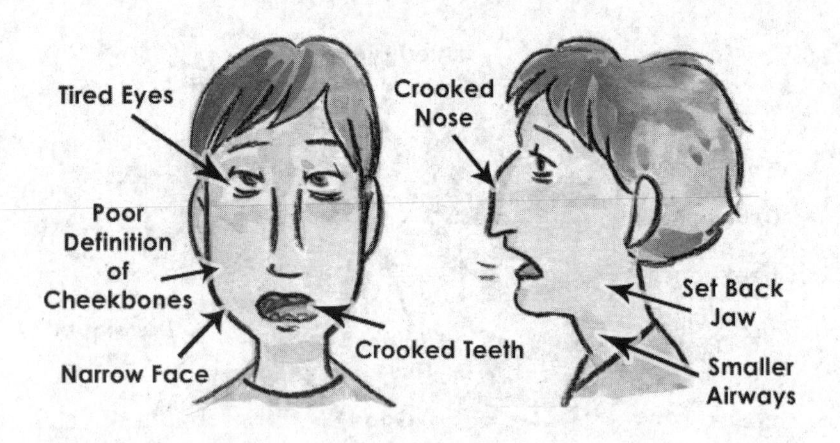

In the image above, the jaws are set back and the airways are smaller, resulting in diminished athletic performance. Had the jaws been in a more forward position, the nose would be straighter and smaller. The eyes look tired as the face sinks downward, and there is poor definition of the cheekbones. Chronic, habitual mouth breathing is also associated with postural changes that result in decreased muscle strength, reduced chest expansion, and impaired breathing. Interestingly, researchers have found that mouth breathers are more likely to be male.

While the above image is exaggerated somewhat, these features are identifiable in thousands of children and adults who have fallen between the cracks of the health care system and were not encouraged to breathe through their nose. These same individuals often suffer from poor health, low energy, and reduced concentration. In the words of Dr. Yosh Jefferson: "These children do not sleep well at night due to obstructed airways; this lack of sleep can adversely affect their growth and academic performance. Many of these children are misdiagnosed with attention deficit disorder (ADD) and hyperactivity."

Dr. Egil Peter Harvold, an expert in orthodontics and craniofacial anomalies, carried out extensive research into the development of the facial structure of monkeys in the 1970s, discovering that the restriction of nasal breathing over several years led to a lowering of the jaws, crooked teeth, and other facial deformities. While today we would consider it dreadful to experiment on innocent animals in this way, hundreds of thousands of children take part in a similar experiment and experience the same craniofacial anomalies due to the effects of mouth breathing. Dr. Harvold's studies paved the way for the treatment and prevention of the improper development of the jaws and face and is accredited almost single-handedly for introducing a branch of orthodontics known as functional appliance therapy to North America.

A 2012 study investigating the long-term changes to facial structure caused by mouth breathing noted that this seemingly "benign" habit "has in fact immediate and/or deferred cascading effects on multiple physiological and behavioral functions." Infants and children who breathe through their mouths due to nasal obstruction are likely to develop crooked teeth and a longer, narrower face, permanently affecting their appearance. Mouth breathing also has a significant impact on the health of the child, including restriction of the lower airways, poor quality of sleep, high stress levels, and a lower quality of life. Research has suggested that habitual mouth breathing may even be connected to sudden infant death syndrome.

Hold On to Your Teeth!

Over the past few years I have been asked to give talks on nasal breathing at dental conferences in Europe, Australia, and the United States. Each conference presents a wonderful opportunity to speak with international experts in dentistry and related disciplines, including orthodontics. There are two specific groups within orthodontics that hold opposing views: functional orthodontists and traditional orthodontists.

A functional orthodontist places emphasis on attaining the correct facial profile of the face as well as straightening the teeth. Functional appliances are worn by the child in order to help guide the growth of the face, jaws, and teeth to reach their full genetic potential. The commonly held view of functional orthodontists is that overcrowding of teeth is not because the teeth are too big, but more likely because the jaw is too small as a result of mouth breathing or thumb sucking. The course of action therefore is to gently expand the jaws and guide them forward to make room for the teeth, with extractions only carried out as a last resort.

In contrast, the primary emphasis of traditional orthodontics is to straighten the teeth, with the facial profile and airway size receiving secondary consideration. Crowded teeth are usually treated by extracting four perfectly good premolars and retracting or pulling the front teeth back to close up the excess space created. This retraction sometimes causes the face, especially around the lips, to look caved in, and the nose and chin to become more prominent. This retraction of the front teeth may also cause problems with the joint of the jaw if the lower jaw is forced back too far. When the lower jaw is too far back, it encroaches on the upper airways, decreasing airway size and negatively affecting athletic ability.

If you are the parent of a child who may be undergoing orthodontic treatment, I would like to impart some advice from international orthodontist Dr. John Mew, who has devoted his life to ensuring the normal growth of children's faces:

- First ask the orthodontist how many permanent teeth will be initially removed and how many more may need extracting later. Sadly, most children who receive traditional retractive orthodontic treatment lose 4 premolar teeth, and nearly half of these children will not have room for their wisdom teeth, leaving them with 24 teeth. This can be completely avoided by helping to ensure normal growth of the face to accommodate all 32 teeth.
- Ask your clinician whether he or she is confident about avoiding an increase in the vertical growth of your child's face. You have the right to be informed of all types of treatment and to be warned of likely problems.

Knowledge is empowering. The best approach to determine the treatment options for you and your child is to research both functional and traditional orthodontic treatment. Selecting the correct approach will have repercussions on the well-being of your child; it is well worth investing a little time to determine a more noninvasive solution. Extractions should be a last resort.

Don't Leave It Until It's Too Late

According to American research, 95 percent of head circumference growth for the average North American white child takes place by the age of nine. Development of the lower jaw, however, continues until approximately the age of eighteen.

Based on these observations, for correct craniofacial growth to take place, early intervention with nasal breathing and tongue posture is essential. The negative effects of mouth breathing on the structure of the jaws and face will have the most impact when they occur before puberty, so there is only a brief window of opportunity to avoid the need for orthodontic treatment and significant changes in a child's facial structure.

My daughter is three years old. She first began to breathe occasionally through her mouth at the age of eight months during teething, and since then I have been encouraging her to breathe through her nose. I set an example by breathing through my own nose all the time, and congratulate her any time she has her mouth closed. As we live in the countryside, I often comment that Charlie the donkey breathes through his nose, as does Snowball the cat. Such clever animals!

Genetically, there would be a high likelihood for my daughter to be a mouth breather as both my wife and I had severe respiratory complaints while growing up. The earlier you begin to encourage your children to breathe through their nose and ensure the correct positioning of the tongue, the better. Not only might you help them avoid orthodontic treatment altogether, but the shape of their face, their general health, and their athleticism will be significantly influenced during these few short years. Even genetic predispositions can be minimized when the right action and behavior is enshrined.

A few years ago I wrote a self-help book for children, teenagers, and parents entitled *Buteyko Meets Dr. Mew: Buteyko Method for Children and Teenagers* about the craniofacial changes associated with mouth breathing. I put extensive amounts of research into the subject and included many peer-reviewed papers and studies to support my claims. Most parents are shocked when they realize that crooked teeth, narrow faces, large noses, and undeveloped jaws can be avoided if a child is simply encouraged to breathe through his or her nose. Not only is sports performance affected, but lifelong health! We cannot be dispassionate about the impact of mouth breathing on the development of our children. The detrimental effects that I suffered from a childhood of breathing through my mouth needn't happen to anyone else, armed with the knowledge we have about the benefits of nasal breathing.

CHAPTER 14

Exercise as if Your Life Depends on It

For two million years we survived without supermarkets, convenience stores, microwaves, or McDonald's. In the past, in order to feed ourselves and our families, muscle movement was required to collect berries and vegetation or to track down an animal—often over several days—until it collapsed from exhaustion.

Compare that to today's couch potato lifestyle—do you know many people who could adequately feed themselves if there was no food available for convenient purchase? One thing is for sure—a major shift of attitude would be required. In order to eat, we would have to perform regular and sustained physical exercise, something that the human body is built for and thrives upon. Why should we stop it now, simply because hunting our own food is no longer necessary?

And while it might be too late for many of us to become elite athletes, we can still enjoy and derive considerable benefit from regular physical exercise. Even if you have lived a relatively sedentary life until now, you can develop your fitness slowly and steadily and bring about positive changes to your health. There is a great feeling of self-esteem and accomplishment when we get up off our backsides or move away from the computer. If you exercise for an hour per day, five times per week, for just a few weeks, you will experience a remarkable difference in your sleep quality, mood, and state of health.

Despite the advances of medical science, the diseases of modern civilization continue to increase, and it could be argued that we are passing responsibility for our health to pharmaceutical companies rather than making changes to our lifestyle and diet to prevent illness in the first place. Each year, more and more people are developing asthma, heart disease, diabetes, high blood pressure, and cancer. Not only that, but new "diseases" are being invented all the time—and some might say it's because the market conditions and share prices of big pharmaceutical companies dictate a need for new markets and ever-increasing sales. We may be living longer, but at the expense of consuming vast quantities of medication as soon as we approach fifty or sixty years of age. A far better option to help prevent the onset of diseases—and one that involves no toxicity or expense—is to take a regular fast walk or jog. If you haven't exercised regularly before, please pay a visit to your physician to get the all clear before starting.

One thing that pretty much all health professionals agree on is the importance of regular physical exercise, taken within our limits. Dozens of studies over the last few years have shown that regular physical exercise provides many health benefits, including a reduced risk of cardiovascular disease, cancer, and diabetes.

Studies as far back as the 1950s have investigated the relationship between regular physical exercise and cardiovascular health. One of the earliest was conducted by Dr. Jeremy Morris, who studied the incidence of heart attacks in 31,000 transport workers. Morris found that bus conductors, who spent most of their day climbing up and down the stairs of double-decker buses, averaging between 500 and 700 steps per day, had reduced incidences of heart disease than their bus-driving counterparts, who spent 90 percent of their day sitting down. Not only this, but heart disease developed by the bus conductors tended to present later in life and was less likely to be fatal.

The same study was replicated using over 100,000 postal workers, and once again it was found that those who spent their day on foot or cycling while delivering the mail had fewer incidences of heart disease compared to workers in sedentary occupations such as telephonists, civil service executives, and clerks. Dr. Morris's findings are just as relevant today as sixty years ago, and since more and more of us spend our working day behind a desk, there is an increased need to perform regular physical exercise. Even more important is that we perform this exercise within our own personal limits.

The best way to ensure that we exercise within our capabilities is by nasal breathing, and an important consideration is the length of time you can comfortably hold your breath, or your BOLT measurement. When your BOLT score is less than 20 seconds, your breathing volume will be greater than what your body requires, resulting in a risk of overbreathing during physical exercise. All breathing should be through the nose when performing physical exercise so as not to increase breathing volume even further.

When your BOLT score is greater than 20 seconds, it's fine for you to breathe through the mouth during exercise for brief periods of time. However, it is only when your BOLT score is greater than 30 seconds that intense physical exercise can be performed without undue risk of overbreathing. The goal for serious athletes, of course, is to attain a BOLT score of 40 seconds.

Only with a high BOLT score of 40 seconds is breathing volume at a normal level. Anything less indicates habitual overbreathing. And while it would be a normal expectation for athletes to possess a high BOLT score, in reality this is seldom the case. I have measured the BOLT scores of hundreds of athletes, ranging from weekend warriors to elite Olympic-level athletes, and guess what? The vast majority of BOLT measurements were less than 20 seconds. Some, in fact, have been less than 10 seconds, as low as any person with severe asthma. It is shocking to think of the stress that such overbreathing is placing on the body during intense physical exercise, so anyone serious about exercise should aim to reduce their breathing and increase their BOLT score.

Like any change to your routine, the best way to determine the benefits of reduced breathing, nasal breathing, and a higher BOLT score is to put the Oxygen Advantage program into practice for 2 to 3 weeks. You will find a program specific to your BOLT score, exercise history, and state of health in Part IV of this book. Two or three weeks is a relatively short period of time in the grand scheme of things, and most individuals will experience positive benefits within a few short days. As you experience the positive benefits, I am in no doubt that you will be happy to incorporate these principles into your life, for the rest of your life.

For example, when I first learned to unblock my nose and switch from mouth to nasal breathing, I immediately felt tension lifting from my head, and despite having wheezed my way through the previous twenty years, my symptoms reduced by as much as 50 percent within the first day, simply by breathing calmly through the nose.

Even if you are unwilling to make a 100 percent change to nasal breathing during your peak season training, bear in mind that whatever you do to incorporate the Oxygen Advantage program into your routine will provide a return. Ultimately, your success depends on your ability to improve your BOLT score—each 5-second improvement will cause the benefits to accumulate.

In Part IV I have outlined a series of different programs to suit individuals according to their current training, BOLT score, age, and overall health. Choose the one that best matches your fitness level and abilities and move on when you feel you have accomplished as much as possible from each program.

Would You Do Me a Favor?

I am very optimistic that if you follow the principles outlined in this book and work toward increasing your BOLT score, your health and sports performance will improve in many ways.

Since 2002, I have observed tremendous improvements in the sporting ability of both amateur and professional athletes who have applied the concepts of the Oxygen Advantage program. I have also witnessed the negative effects on respiratory, mental, and cardiovascular health from chronic overbreathing. It is my aim to spread the word of these simple techniques to provide an easy and effective way of improving the health and well-being of readers.

And you can help me to generate awareness!

I would appreciate it greatly if you would try out these practices and then spread the word to your friends and family who are interested in health and sports, or who suffer from any of the problems described in this book. You could also help by writing a review of *The Oxygen Advantage* on Amazon.com.

Also, please share the book with anyone you think may benefit from the techniques within the Oxygen Advantage program.

Finally, if you have any questions, or would simply like to tell me about your personal experience with the Oxygen Advantage program, please drop me a line. I would love to hear from you. You can contact me directly at patrick@OxygenAdvantage.com.

Best wishes and thanks again,
Patrick McKeown
Galway, Ireland

PART IV

Your Oxygen Advantage Program

Summary and General Program Based on BOLT Score and Health

Each time I work with a client, I design a program of breathing exercises and lifestyle guidelines to help him or her reach goals safely and in the shortest time possible. When deciding on a program, it is necessary to take into consideration each individual's health and BOLT score. Having an insight into a person's lifestyle is also useful, as exercises can be tailored to cause minimal disruption to work schedules and current training routines. I fully understand the challenge of making time in a busy life to exercise, which is why the Oxygen Advantage program offers quick, easy, and achievable techniques that can fit around any existing routine. There is no doubt that a low BOLT score can result in fatigue, reduced concentration, and poor productivity, so by allocating just a half hour to one hour each day to practice these exercises, your BOLT score will increase along with your energy levels, well-being, and performance. I have seen the proof in thousands of clients that spending a little time to improve body oxygenation is a hugely beneficial investment.

The best way to approach the Oxygen Advantage program is to view it as a lifestyle change and something to be incorporated into your way of life as opposed to a set of exercises practiced formally throughout the day. This way it will become part of your daily routine rather than a regular chore or duty.

Quick Reference Summary of the Oxygen Advantage Program

Habitual overbreathing involves breathing more air than your body requires during rest and exercise. Overbreathing leads to:

- A reduction of the gas carbon dioxide in the blood
- Mouth breathing and underutilization of the gas nitric oxide
- Impaired release of oxygen from red blood cells (see the Bohr Effect, page 26)
- Constriction of the smooth muscle in the blood vessels and airways
- Adverse effects on blood pH
- Reduced oxygenation of working muscles and organs, including the heart and brain
- Increased acidity and fatigue during exercise
- Limited sports performance
- Negative effects to overall health

Benefits of practicing the Oxygen Advantage program include:

- Improved sleep and energy
- Easier breathing with reduced breathlessness during exercise
- Naturally increasing the production of EPO and red blood cells
- Improving oxygenation of working muscles and organs
- Reduction of lactic acid buildup and fatigue
- Improved running economy and VO_2 max
- Improved aerobic performance
- Improved anaerobic performance

Oxygen Advantage Exercise Summary

The following BOLT measurement and exercises are explained in detail in earlier chapters.

Body Oxygen Level Test (BOLT)

1. Nose Unblocking Exercise
2. Breathe Light to Breathe Right
3. Breathe Light to Breathe Right—Jogging, Running, or Any Other Activity
4. Breathing Recovery, Improved Concentration
5. Simulate High-Altitude Training—Walking
6. Simulate High-Altitude Training—Running, Cycling, Swimming
7. Advanced Simulation of High-Altitude Training

Body Oxygen Level Test (BOLT) (see page 37 for more detail)

Your progress can be determined by a reduction to breathlessness during physical exercise, how you feel, and your **BOLT score** as follows:

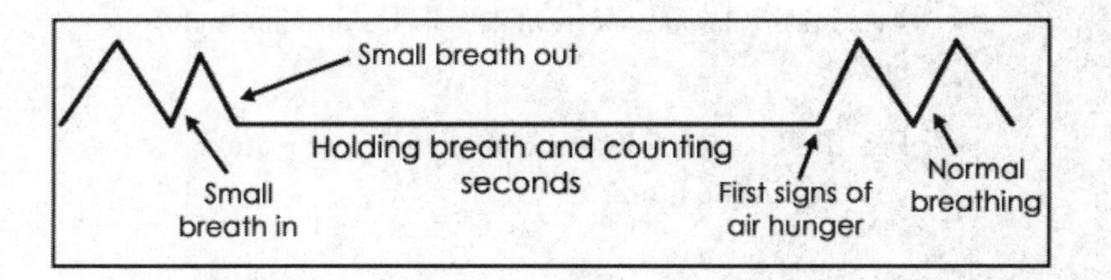

1. Take a small, silent breath in through your nose, and allow a small, silent breath out through your nose.
2. Hold your nose with your fingers to prevent air from entering your lungs.
3. Count the number of seconds until you feel the first definite desire to breathe.
4. At the first definite desire to breathe in, you may also feel the first involuntary movements of your breathing muscles. (Your abdomen may jerk and the area around your neck may contract.)
5. Release your nose and breathe in through it.
6. Your inhalation at the end of the breath should be calm.

Your **BOLT** score is the length of time in seconds that you are able to hold your breath until you feel the first physical signals to take a breath. To increase your BOLT score, it is necessary to:

- Nose-breathe at all times, including during physical exercise and sleep.
- Avoid taking large breaths while sighing, yawning, and talking.
- Practice Oxygen Advantage exercises appropriate to your health and fitness.

By incorporating the Oxygen Advantage exercises into your routine, your BOLT score should increase by 3 to 4 seconds during the first week. After continued practice for a number of weeks, you may find your BOLT score remains stubbornly at about 20 seconds. Continue to practice and incorporate breath holds into your physical exercise to increase your BOLT score above 20 seconds. It can take 6 months for an individual to reach a BOLT score of 40 seconds, but by then your health and fitness levels will be at a totally different place than before. Enjoy the journey!

1. Nose Unblocking Exercise (see page 61 for more detail)

(Please do not practice this exercise if your BOLT score is less than 10 seconds, or if you are pregnant or have high blood pressure, cardiovascular issues, diabetes, or any serious health concerns.)

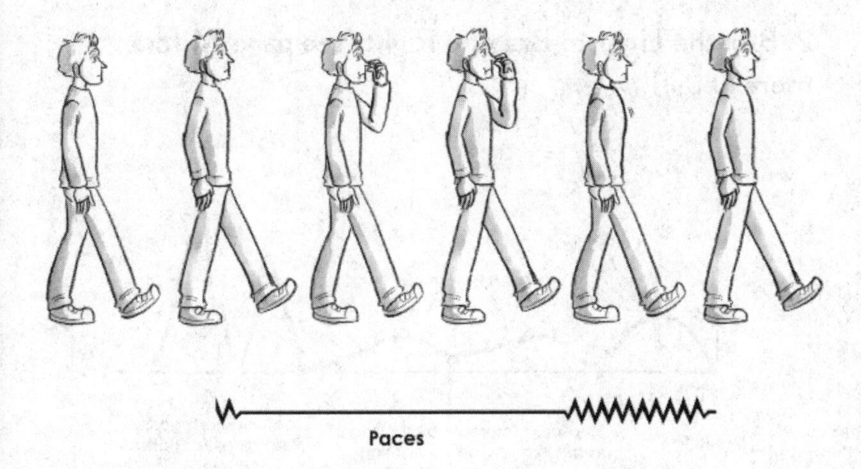

To unblock the nose, perform the following:

1. Take a small, silent breath in and a small, silent breath out through your nose.
2. Pinch your nose with your fingers to hold your breath.
3. Walk as many paces as possible with your breath held. Try to build up a strong air shortage, without overdoing it, of course!
4. When you resume breathing, do so only through your nose; your breathing must be calmed immediately.
5. After resuming your breathing, your first breath will usually be bigger than normal. Make sure that you calm your breathing as soon as possible by suppressing your second and third breaths.
6. You should be able to recover your breath within 2 to 3 breaths. If you cannot, you have held your breath for too long.
7. Wait for about a minute or so and then repeat.
8. Repeat this exercise 5 or 6 times until the nose is decongested.

2. Breathe Light to Breathe Right (see page 74 for more detail)

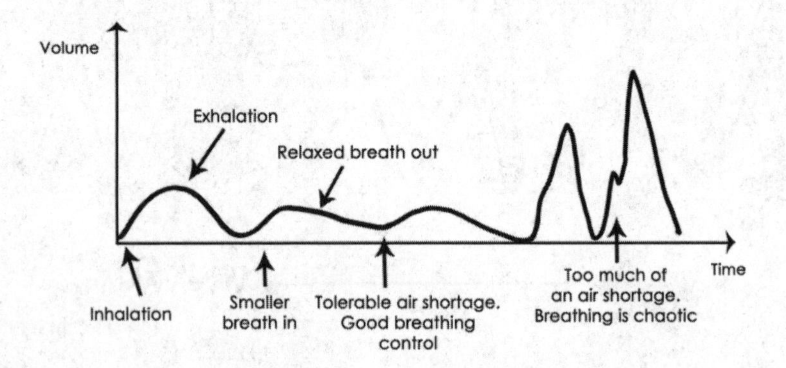

1. Place one hand on your chest and the other just above your navel to help you to follow your breathing.
2. Breathe in and gently guide your abdomen outward.
3. Breathe out and gently guide your abdomen inward.
4. Observe your breathing pattern, noting the size and depth of each breath.
5. Apply gentle pressure with your hands to slightly reduce your breathing movements. It should feel as if you are breathing against your hands.
6. Encourage the depth of each breath to reduce.
7. Take in a smaller or shorter breath than you would like.
8. Allow a relaxed breath out, exhaling gently, slowly, and easily.
9. Bring a feeling of relaxation to your breathing.
10. Do not tense your body, hold your breath, or pause your breathing. Continue to breathe smoothly but take in less air than before.

11. The objective of this exercise is to create a tolerable hunger for air. Try to sustain this for 3 to 5 minutes at a time. If your breathing rhythm becomes chaotic or if your breathing muscles contract, then the air shortage you have created is too much. If these signs occur, stop the exercise and return to it when breathing is back to normal.

3. Breathe Light to Breathe Right—Jogging, Running, or Any Other Activity (see page 89 for more detail)

No matter what type of exercise you prefer, make sure that you observe your breath and become aware of your inner body. Bring your entire attention from your mind into your body. Move with every cell of your body, from the top of your head to the tips of your toes.

Allow your body to find its perfect work rate by breathing through your nose in a steady and regular fashion. Continue to increase your pace to the point where you can maintain steady and regular nasal breathing. If your breathing rhythm becomes chaotic and it is necessary to open your mouth to breathe, you will know that the intensity is too much. If necessary, slow your pace to a walk for 2 to 3 minutes before resuming your jog.

As you run, feel each gentle connection between your feet and the ground as you propel yourself forward. Avoid pounding the pavement as this will lead to sore hips and joints and other possible injuries. Instead, bring a feeling of lightness to your body and visualize yourself barely touching the ground as your run. Imagine yourself running over thin twigs, treading so softly that they do not break. The mantra to follow is: light foot strikes, a relaxed body, and regular, steady breathing.

If you keep your mouth closed throughout your exercise, your breathing will recover quickly.

4. Breathing Recovery, Improved Concentration (see page 91 for more detail)

Small breath in | Small breath out | Hold breath for 2–5 seconds | Breathe normally for 10–15 seconds | Continue until calm

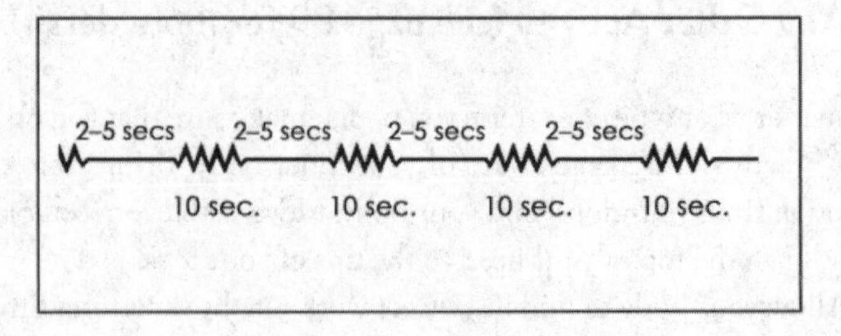

To recover from physical exercise and to help calm your breathing or mind, practice the following exercise for 3 to 5 minutes:

1. Exhale as normal through the nose.
2. Pinch your nose with your fingers to hold the breath for 2 to 5 seconds.
3. Breathe normally through the nose for 10 seconds.
4. Repeat the first three steps.

Important Note Regarding the "Simulate High-Altitude Training" Exercises That Follow

Please do not practice any exercise that simulates high-altitude training if your BOLT score is less than 20 seconds (or less than 30 seconds for advanced simulation of high-altitude training) or if you are pregnant, have high blood pressure or other cardiovascular issues, have diabetes, or have any serious health concerns. While these exercises involve the creation of a medium to strong air hunger, there is no point in overdoing it. Upon completion of each breath hold, you should be able to recover your breathing within 2 to 3 breaths. If while practicing these exercises, you experience dizziness or any other negative side effect, then please stop immediately.

5. Simulate High-Altitude Training—Walking (see page 126 for more detail)

If you have a pulse oximeter, you might find it motivating to observe the decrease to your oxygen saturation as you do this. Continue walking throughout the exercise and hold your breath only until you feel a medium hunger for air for the first 2 to 3 breath holds. For the remaining breath holds, it is beneficial to hold the breath until you feel a relatively strong hunger for air.

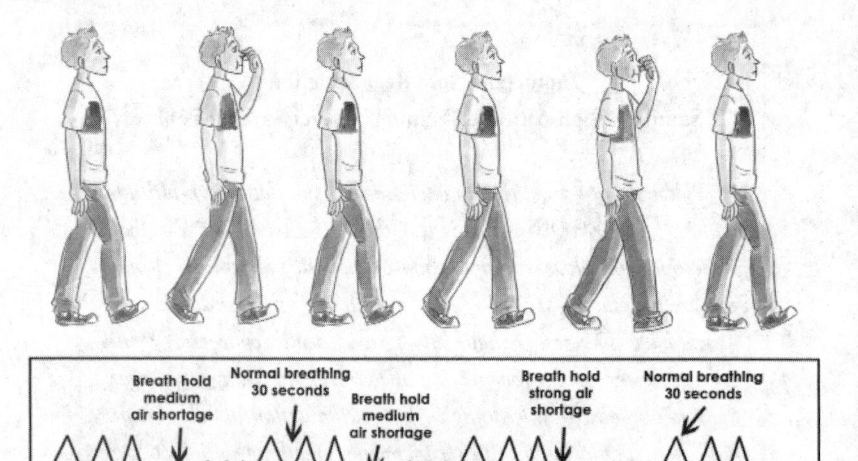

1. Walk for 1 minute or so while breathing through your nose.
2. **Gently exhale and hold your breath, followed by minimal breathing for 15 seconds:** Gently exhale, pinch your nose, and walk while holding the breath until you feel a medium hunger for air, then release your nose, inhale through it, and minimize your breathing for 15 seconds by taking short breaths. After 30 seconds of continued walking and nose breathing, repeat the breath hold until you feel a medium hunger for air. Minimize your breathing for 15 seconds, then allow your breathing to return to normal and through the nose.
3. **Continue walking for 30 seconds and repeat:** Continue walking for around 30 seconds while breathing through your nose, then gently exhale and pinch your nose with your fingers. Walk while holding your breath until you feel a medium to strong hunger for air. Release your nose and minimize your breathing by taking short breaths for about 15 seconds. Then resume breathing through your nose.

4. **Repeat breath holds 8 to 10 times:** While continuing to walk, perform a breath hold every minute or so in order to create a medium to strong need for air. Minimize your breathing for 15 seconds following each breath hold. Repeat for a total of 8 to 10 breath holds during your walk.

A typical increase in the number of paces per breath hold might look like this: 20, 20, 30, 35, 42, 47, 53, 60, 60, 55.

6. Simulate High-Altitude Training—Running, Cycling, Swimming (see pages 130–132 for more detail)

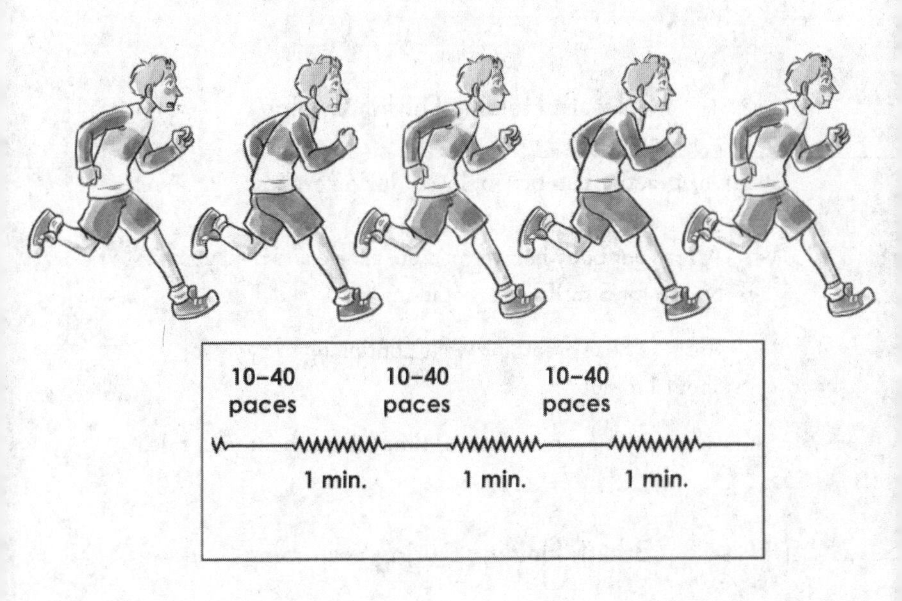

Breath Holding During Running

Breath holding can also be incorporated into more intense exercise, such as running:

1. 10 to 15 minutes into your run, gently exhale and hold your breath until a strong air shortage is reached. The length of the breath hold may range from 10 to 40 paces and will depend on your running speed and BOLT score.
2. Following the breath hold, continue to jog with nose breathing for about 1 minute, until your breathing has partially recovered.
3. Repeat the breath hold 8 to 10 times for the duration of your run. The breath hold should be a challenge and, at the same time, should allow breathing to recover to normal within a couple of breaths.

Breath Holding During Cycling

A similar practice can be employed during cycling:

- After your body has warmed up, exhale and hold your breath for 5 to 15 pedal rotations.
- Resume nose breathing while continuing to cycle for about 1 minute.
- Repeat this exercise 8 to 10 times throughout your ride.

Breath Holding During Swimming

During swimming, increase the number of strokes between breaths. You can do this in gradual increments, increasing the number of strokes between breaths from 3 to 5 to 7 over a series of lengths.

7. Advanced Simulation of High-Altitude Training (see page 132 for more detail)

For this exercise, it is necessary to monitor your blood oxygen saturation with a pulse oximeter, ensuring that your SpO_2 does not drop below 80 percent.

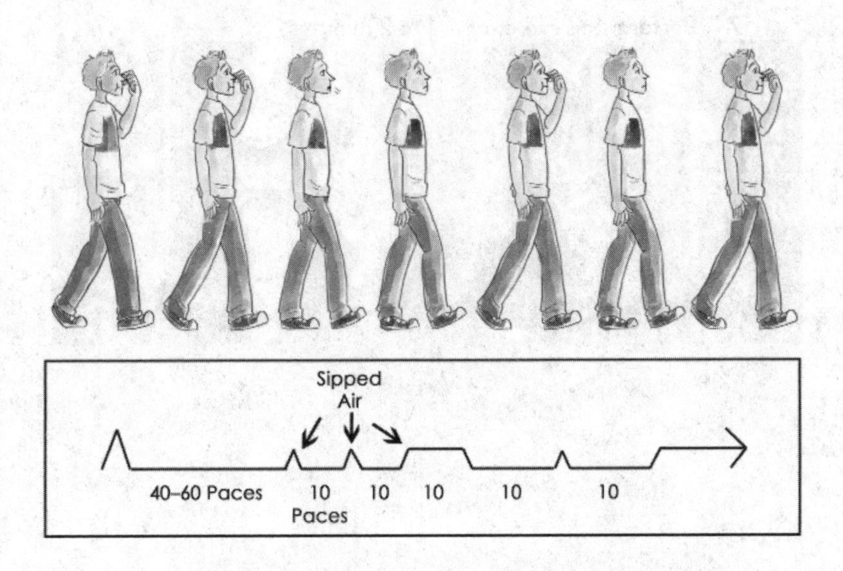

1. Walk for a minute or so. Exhale and hold your breath for approximately 40 paces, then take a sip of air into the lungs. A "sip" is a tiny breath—just enough to reduce tension. Hold the breath for a further 10 paces.

2. Now take a sip of air in or out of the lungs. Hold for 10 paces or so.

3. Continue taking sips of air and repeating short breath holds until you feel a fairly strong air shortage.

4. If the air shortage is too strong, then reduce the hold to 5 paces or less. With each successive breath hold, oxygen saturation will continue to decrease.

5. Challenge but do not stress yourself.

6. Continue to monitor pulse oxygen saturation. Do not go below 80 percent SpO_2.

7. Perform this exercise for 1 to 2 minutes.

Breathe Light to Breathe Right (Advanced Method)

The advanced version of Breathe Light to Breathe Right is best practiced after you have mastered the basic exercise as described on page 74. The following will teach you how to combine reduced breathing with abdominal breathing to increase your BOLT score.

Please note that just like any form of physical exercise, it is easier and more beneficial if you practice reduced breathing exercises at least an hour or more after eating a meal.

This exercise involves three simple stages:

1. Activating and strengthening the diaphragm
2. Merging your breathing with the movements of your abdomen
3. Reducing your breathing to create an air shortage

By practicing abdominal breathing, you will allow it to become your natural breathing method. This exercise is split into three parts to ensure that you learn the technique correctly and to gradually incorporate abdominal breathing into your normal breathing habits. First you must learn to relax the diaphragm muscle so that it can be activated during breathing. Second, you must learn to match the movements of the abdomen with the breath to engage the diaphragm. And finally, you will be able to practice breathing lightly with abdominal breathing to maximize body oxygenation during rest. Remember that to improve your breathing during exercise, you must first learn how to breathe efficiently at rest.

Stage 1: Relaxing and Activating the Diaphragm

- Sit up straight but do not force yourself into a rigid upright position, as doing so will only increase the tension in your body. Instead, try to lengthen the distance between your navel and sternum (chest); imagine a piece of string gently lifting you up from the top of the back of your head.
- As you guide your body upward, imagine the space between your ribs becoming wider.
- Place one hand on your chest and one hand just above your navel. At this point, do not concern yourself with how you are breathing.
- Bring your attention to the movements of your lower hand. While sitting up straight, gently guide your hand outward by pushing your abdomen outward, just enough to feel the movement. There is no need to make any changes to your breathing at this point—this stage is primarily to encourage abdominal movement.
- Now draw in your abdomen and watch your hand move gently inward.
- Perform this simple exercise for a few minutes to help activate a "stiff" diaphragm.
- Alternatively, you could perform this exercise while lying on your back with both knees bent and your feet flat on the floor.

Stage 1 in Brief

- Gently push your abdomen out. Watch your hand move outward.
- Gently draw your abdomen in. Watch your hand move inward.

If your diaphragm is particularly stiff due to years of breathing with the upper chest, the following additional exercise will help to activate the muscle and encourage better breathing:

- Take a gentle breath in through your nose.
- Allow a gentle breath out through your nose.
- Hold your nose with your fingers and close your mouth to prevent air from flowing.
- Now try to breathe in and out, while at the same time holding your breath.
- As you try to breathe in and out, you may feel your stomach move in and out as your breathing muscles contract to help relax the diaphragm.
- When you feel a medium desire to breathe, let go of your nose and resume breathing normally through your nose.
- Practice this exercise 2 or 3 times to help to relax the diaphragm.

When you feel you can move your abdomen in and out easily at will, proceed to **Stage 2**, which incorporates these abdomen movements with your breathing.

Stage 2: Merging Abdominal Movements with Breathing

- Sit up straight.
- Place one hand on your chest and the other hand on your abdomen.
- As you breathe, allow your shoulders to relax into their natural position.
- Gently encourage chest movements to reduce as you breathe, using the guidance of your mind and your hand.
- At the same time, try to coordinate your abdomen movements with your breathing.
- As you breathe in, gently guide your abdomen outward. Imagine you are breathing into your tummy. (Try not to make the movement too big, as doing so may cause dizziness.)
- As you breathe out, gently draw your abdomen in.
- Your breathing should be gentle, silent, and calm.
- Perform this exercise for a few minutes to acclimatize yourself to the movements of your diaphragm and your breath.

If you are finding this exercise difficult, you may find it easier to activate abdominal breathing while lying on your back in a semi-supine position. Try the following exercise lying on a mat with a small pillow under your head and your knees bent, as shown on the following page.

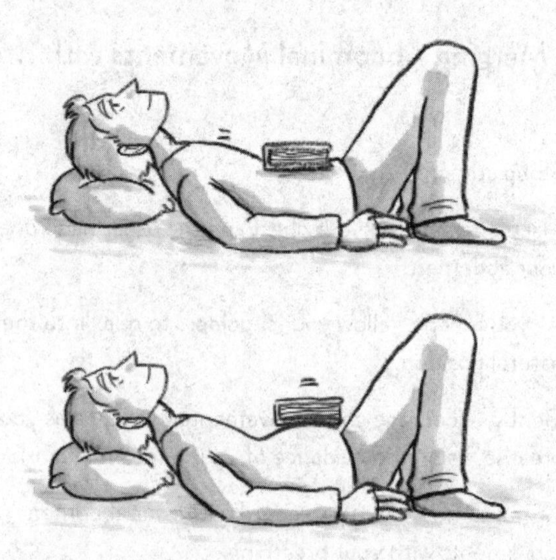

- Place a relatively large book on the area just above your navel.
- As you breathe in, bring your breathing into your abdomen by gently guiding your abdomen to move the book upward.
- As you breathe out, gently allow the abdomen to move back to its original position.
- The inhalation is the active phase, and the exhalation is passive, as you allow the air to leave the body naturally and effortlessly. While breathing in, imagine inflating your tummy with a light amount of air and watch the book rise. While breathing out, imagine a balloon slowly deflating of its own accord.

Stage 2 in Brief

- Breathe in. Gently guide your abdomen out.
- Breathe out. Gently draw your abdomen in.

When you are confident that you are able to match your breathing with the movements of your diaphragm, proceed to **Stage 3**.

Stage 3: Reducing Breathing Volume Using Abdominal Breathing

If you have tried **Stages 1** and **2** but still find you are unable to switch from upper-chest to abdominal breathing, don't be concerned—it can take time to readjust to a new way of breathing after years of breathing from the upper chest. You can still proceed to **Stage 3**—simply keep practicing all three stages until it becomes easier. The more you use these exercises, the greater your tolerance of carbon dioxide will become.

Reducing your breathing is about lowering the amount of air you take into your lungs during each minute. As you reduce your breathing, there will be a slight accumulation of carbon dioxide in the blood, assisting with the relaxation of the diaphragm. If you have already mastered diaphragmatic breathing through the first two stages, then you will find that it is a lot easier to reduce your breathing volume toward normal as you practice **Stage 3**.

There are two main approaches to reducing breathing volume using diaphragmatic breathing. The first is to bring a feeling of relaxation to the body, to gently allow your breathing movements to slow down and become softer. As your body relaxes, your breathing will automatically reduce in response. The second approach is to tune in to your breathing pattern, noting the size of each inhalation and exhalation. By concentrating on the rhythm of your breathing for a minute or two, you will gain a sense of how much air you are taking into your body. As you follow your breathing, gently encourage it to slow down so that your breathing movements gradually reduce to the point where you feel a subtle need for air.

This urge to breathe is the core of reduced breathing exercises and is a sign that you are actively altering your breathing habits toward a healthier, more effective level. When you first attempt reduced breathing exercises, you may find the sensation of air shortage difficult to maintain, but it is essential to keep practicing if you want to bring about changes to your body and improve your sporting performance. The following sentence is the most important point of this entire book, and something I explain to my students daily:

> **The only way to know you are reducing your breathing volume is feeling as if you would like to take in a bigger breath.**

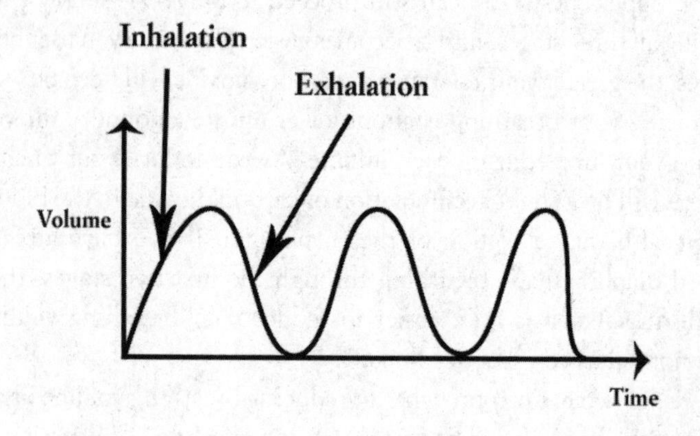

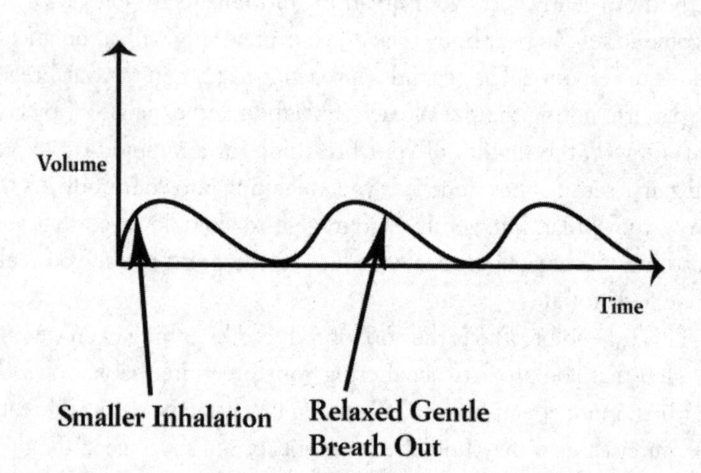

The need to take a bigger breath is similar to the feeling you will have experienced during your measurement of BOLT. This urge for air should not be stressful, but should be similar to what you might experience during a normal walk.

In **Stage 3,** we will bring abdominal breathing and reduced breathing together. To practice this exercise, it can be very helpful to sit in front of a mirror to observe and follow your breathing movements:

- Sit up straight.
- Place one hand on your chest and one hand on your abdomen.
- Imagine a piece of string guiding you upward from the top of the back of your head. Imagine the space between your ribs gently widening.
- As you breathe in, gently guide your abdomen out. Keep your chest movements to a minimum.
- As you breathe out, gently guide your abdomen in, still keeping your chest movements small.
- Follow each breath in and out through your nose.
- Tune in to the amplitude of each inhalation and exhalation. Try to get a sense of the size and frequency of each breath.
- As you breathe, exert gentle pressure with your hands against your abdomen and chest. This should create extra resistance to your breathing.
- Breathe against your hands, concentrating on making the size of each breath smaller.
- With each breath, take in less air than you would like to. Make the in-breath smaller or shorter.

- Breathe out with a relaxed exhalation. Allow the natural elasticity of your lungs and diaphragm to play their role in each exhalation. Imagine a balloon slowly and gently deflating of its own accord.
- When the in-breath becomes smaller and the out-breath is relaxed, visible breathing movements will be reduced. You may be able to notice this in a mirror.

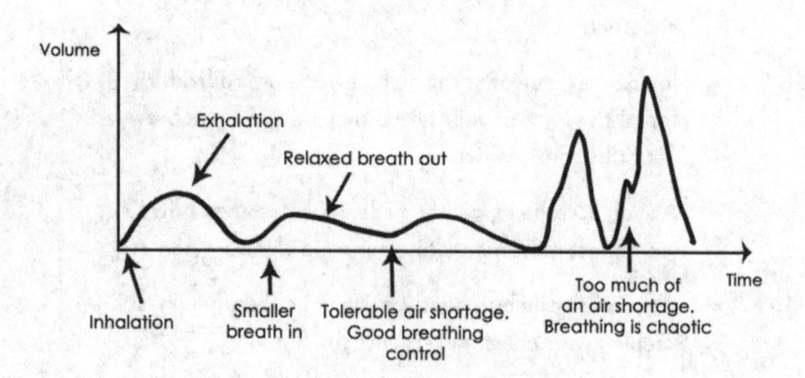

By using a simple exercise like this you can reduce your breathing movements by 20 to 30 percent. If your stomach muscles start to feel tense, contract, or jerk, or if your breathing rhythm becomes disrupted or out of control, then the air shortage is too intense. In this situation, abandon the exercise for 15 seconds or so and return to it when the air shortage has disappeared.

The most common mistake is to deliberately tense the muscles of the chest or abdomen to restrict breathing movements. If you find this happening, then take a break from the exercise for 15 seconds or so. When you return to it, encourage your breathing to reduce by exerting gentle pressure on your chest and abdomen with your hands, encouraging your breathing to slow and diminish using relaxation rather than force.

Do not be concerned about the number of breaths you take per minute. Ideally, this should not increase. However, if your BOLT score is less than 20 seconds, you may find that your breathing rate increases during the exercise. If this happens, try to slow down your breathing and keep it calm. As your BOLT score increases it will become much easier to maintain control of your breathing during reduced breathing exercises.

At first, you may only be able to maintain an air shortage for 20 seconds before the urge to breathe is too strong. With practice, you will be able to maintain an air shortage for longer periods. Remember, you are trying to create an air shortage that is tolerable but not stressful. Aim to maintain this tolerable air hunger for 3 to 5 minutes at a time. Practicing 2 sets of 5-minute exercises is enough to help you reset your breathing center and improve your body's tolerance of carbon dioxide.

During reduced breathing exercises it is vital that you create a hunger for air in order to bring about an accumulation of carbon dioxide in the blood. When this happens, the respiratory center in the brain is reset toward a calmer and more normal breathing volume. In order to reset the respiratory center by just a little, it is necessary to experience an air hunger for about 10 minutes. You can divide most of the exercises in this book into 2 sets of 5-minute sessions, or, if you feel you are confident and experienced in reduced breathing, you may practice for 10 minutes straight.

The Oxygen Advantage: Encompassing Many Factors to Improve Health and Fitness

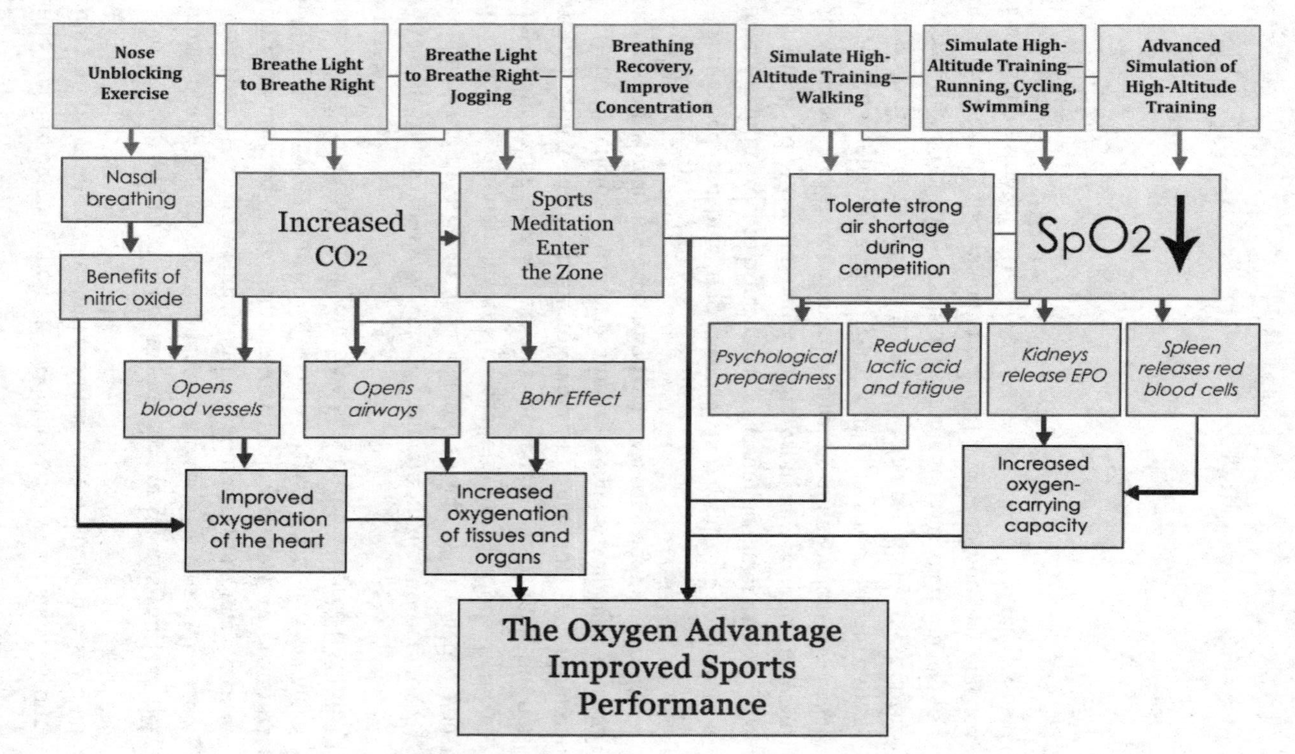

General Program Based on BOLT and Health

Please note: Oxygen Advantage exercises that involve holding the breath to create a medium to strong air shortage during walking, jogging, or running create a similar effect to high-intensity training. As a result, they are not suitable if you are elderly, pregnant, or have high blood pressure, cardiovascular disease, type 1 diabetes, kidney disease, depression, cancer, or any serious health concern. Instead, practice nasal breathing and the gentle Breathe Light to Breathe Right exercise (see page 74) until these conditions are resolved.

Similar to physical exercise, it is recommended to practice the Oxygen Advantage exercises at least two hours after eating.

Program for BOLT Score of Less Than 10 Seconds (or an Unhealthy or Older Person)

- Measure your BOLT score each morning after waking.
- Breathe through the nose both day and night. To ensure nasal breathing at night, it will be necessary to wear paper tape across your lips during sleep (see page 64).
- Practice the Breathing Recovery Exercise (page 91) throughout the day, ideally spending 10 minutes 6 times per day doing small breath holds of between 2 and 5 seconds.
- Another option to help recover breathing is to exhale through your nose, pinch your nose with your fingers, and walk while holding the breath for 5 to 10 paces. Rest for 1 minute and repeat 10 times (see pages 219–220).
- Engage in 10 to 15 minutes of slow walking each day with the mouth closed. If you need to breathe through your mouth, you must stop walking to recover your breath.
- When your BOLT score increases to 15 seconds, you will find it a lot easier to bring relaxation to your body and to Breathe Light to Breathe Right. It is more beneficial to practice this exercise than Breathing Recovery once your BOLT score has reached 15 seconds. The minimum

time required for an individual for such a low BOLT score is to practice Breathing Light to Breathe Right for 1 hour per day (6 sets of 10 minutes each).

- As your BOLT score increases, it will become a lot easier to engage in physical exercise. Your expected progress is to increase your BOLT score to 25 seconds within 6 to 8 weeks.
- Fill in the following chart to mark your progress.

BOLT score of less than 10 seconds	Example	Day 1	Day 2	Day 3	Day 4	Day 5	Day 6	Day 7
BOLT	7 A.M. 7 seconds							
Breathing Recovery	7 A.M. 10 minutes							
Breathing Recovery	10 A.M. 10 minutes							
Breathing Recovery	11 A.M. 10 minutes							
Breathing Recovery	2 P.M. 10 minutes							
Breathing Recovery	3 P.M. 10 minutes							
Breathing Recovery	9 P.M. 10 minutes							
Slow Walk	4 P.M. 10 minutes							

Sixty-five-year-old Michael enjoys slow walking. He has chronic asthma and suffers from symptoms such as coughing, breathlessness, and wheezing. Michael's BOLT is 7 seconds. This program emphasizes quality over quantity.

At the beginning, given Michael's low BOLT score, he will find it difficult to Breathe Light to Breathe Right because the air hunger may destabilize his breathing. Therefore, a better initial option for him is to concentrate on breathing only through the nose both during the day and at night, and to practice many small breath holds (Breathing Recovery Exercise) throughout the day. It will also be very helpful for Michael to relax his body and encourage his breathing to soften, as long as doing so is not stressful and does not destabilize his breathing.

Gentle walking alone will be sufficient for Michael to generate a tolerable air shortage. He will not need to hold his breath during his walk unless he feels comfortable doing so. Sometimes, holding the breath during a walk is a good way to overcome chest tightness and to help increase one's BOLT score. If Michael practices breath holds during his slow walks, they should be limited to no more than 10 paces. He should not hold his breath beyond the point at which he loses control of his breathing, as doing so will disrupt his breathing and possibly cause asthma symptoms.

Practicing the exercise Breathe Light to Breathe Right is very worthwhile and will be a lot easier for Michael to apply when his BOLT score reaches 15 seconds. Michael's BOLT score will continue to increase as long as he dedicates 10 minutes by 6 times daily to his reduced breathing. This may sound like a lot of hard work, but living with asthma is extremely hard work in itself and will already be costing Michael a lot in terms of reduced quality of life and productivity. Reducing his breathing throughout the day will be the best investment in time that Michael has ever made.

Program for BOLT Score of 10 to 20 Seconds

- Measure your BOLT score each morning after waking.
- Breathe through the nose at all times. Wear paper tape across your lips during sleep to ensure nasal breathing at night (see page 64).
- Regularly observe your breathing throughout the day to ensure it stays calm and soft.
- Swallow or hold the breath any time you feel a sigh coming. If you miss a sigh, then gently exhale through your nose and hold your breath for 5 to 10 seconds to compensate.
- Practice the Breathe Light to Breathe Right (page 74) or Breathing Recovery (page 91) exercise for 10 minutes, 3 times a day: once in the morning, once during the afternoon, and once before bed.
- Practice Breathe Light to Breathe Right—Walking (chapter 3) for between 30 and 60 minutes per day; you can slow jog if your BOLT score is greater than 15 seconds.
- Fill in the following chart to mark your progress.

BOLT score of 10 to 20 Seconds	Example	Day 1	Day 2	Day 3	Day 4	Day 5	Day 6	Day 7
BOLT	6:30 A.M. 15 seconds							
Breathe Light	6:30 A.M. 10 minutes							
Breathe Light	8 A.M. 10 minutes							
Breathe Light	10 P.M. 10 minutes							
½ hr–1 hr Exercise	3 P.M. 40 minutes							

Jennifer works as a sales and marketing manager for a UK-based clothing store. Her work can be demanding, resulting in long hours spent in her car or in front of the computer. Travel takes up a large amount of her time, both to and from the head office and from store to store. Because of her busy schedule, Jennifer has gotten out of the habit of regular physical exercise. As she approaches her thirty-fifth birthday, she has become more conscious of her health and fitness and has decided to embark on a new exercise program.

Jennifer's starting BOLT score was 12 seconds, with no apparent health concerns. In the beginning, it was important to avoid the common mistake of overdoing it. In an effort to make up for years of not exercising, there is often a temptation to plunge straight into intensive exercise, but performing beyond your ability can lead to increased breathlessness and feelings of failure. This can sometimes be enough to put you off a new program completely when you would have made much better progress by simply taking exercises at your own pace. The mantra for beginning any exercise program should always be: slow and steady, increasing the intensity and duration by no more than 10 percent each week.

I started Jennifer off very gently, matching specific exercises to her BOLT score and fitness level. To gently condition her body to a greater tolerance of carbon dioxide, Jennifer increased her daily walks to a light jog as soon as her BOLT score reached 20 seconds. During her first week of jogging, she alternated between 2 minutes of walking and 2 minutes of jogging. In the second and third weeks, she began jogging for 3 minutes followed by walking for 1 minute. By the fourth week she had achieved her long-awaited goal of jogging for a half hour straight, and more important she was able to do this comfortably with her mouth closed. All in all, Jennifer's commitment to nasal breathing, gentle reduced breathing exercises, and regular physical exercise achieved impressive benefits with no risk of injury or feeling defeated from overtraining. By intelligently adapting a new exercise program to her needs, Jennifer was able to enjoy her newfound fitness in a way that fit easily around her work and routine.

Program for BOLT Score of 20 to 30 Seconds

- Measure your BOLT score each morning after waking.
- Breathe through the nose both day and at night, including wearing tape over the mouth during sleep (page 64).
- Reduce breathing using the Breathe Light to Breathe Right exercise (page 74) for 10 minutes, 3 times per day, once in the morning, once in the afternoon, and last thing at night.
- Warm up for 10 minutes by walking and performing a breath hold to achieve a medium to strong air hunger every minute or so to Simulate High-Altitude Training (page 126).
- Breathe Light to Breathe Right during a fast walk or jog for 30 to 60 minutes daily with a relaxed body, abdominal breathing, and nasal breathing to create an air shortage.
- Simulate High-Altitude Training during walking or jogging by practicing 8 to 10 breath holds.
- After physical exercise, practice the Breathing Recovery Exercise (page 91).
- Fill in the following chart to mark your progress.

BOLT score of 20 to 30 seconds	Example	Day 1	Day 2	Day 3	Day 4	Day 5	Day 6	Day 7
BOLT	6:15 A.M. 25 seconds							
Breathe Light	6:15 A.M. 10 minutes							
Breathe Light	10 A.M. 10 minutes							
Breathe Light	10 P.M. 10 minutes							
Breathe Light Walk or Jog	3 P.M. 45 minutes							
Simulate High-Altitude Training	Completed during physical exercise above							

David is twenty-three years old and a keen athlete. He trains four times per week and plays rugby for his hometown. His BOLT score is 20 seconds.

David has been mouth breathing and sighing regularly for several years. He snores during his sleep and wakes up most mornings with a dry mouth, nasal congestion, and fatigue. He also finds that during training his breathing is noisier than that of his teammates (who uncharitably call him "the train"). David's heavy breathing can be heard long before he makes a tackle, with his opponents receiving advance warning that he is in pursuit. When I first met David I took note of his narrow facial structure, flaccid lower lip, and slightly protruding and crooked nose, indicating that he had been breathing through his mouth since childhood.

Like many athletes who have invested many years training to maintain their fitness, David was at first a little reluctant to make a drastic change to his regimen. To help dispel any fears, I sat him down and we discussed basic physiology, the importance of optimal breathing for sports success, the theory behind BOLT measurement, and the benefits of simulating high-altitude training.

David is able to maintain breathing through his nose while jogging lightly. However, during intense training, he finds doing nasal breathing too difficult. Training using nasal breathing creates resistance to breathing simply because the nose is a smaller area to breathe through than the mouth. David's fear is that he may lose muscle conditioning because of the decrease in training intensity resulting from having to keep his mouth closed. Taking this situation into account, the best approach for David is to apply 90 percent of the Oxygen Advantage program in order to improve his BOLT score. This includes simulating a high altitude during jogging, breathing through the nose at all times, and reducing his breathing by relaxing during both rest and regular training. The only exception to nasal breathing should be when training becomes so intense that he needs to periodically breathe through his mouth. In time, as David's BOLT score increases, he will be able to maintain nasal breathing during high-intensity exercise. In the meantime, he is able to check whether he is overbreathing during training by comparing his BOLT score before training and 1 hour after completing his exercise. His BOLT score after training should be approximately 25 percent higher than before. If it is lower, David should reduce the intensity of his training to the point at which he can maintain strictly nasal breathing.

David's goal is to reach a BOLT score of 40 seconds within 12 weeks. Physical training with a larger air hunger is the key to achieving this goal.

Program for BOLT Score of 30 Seconds Plus

- Measure your BOLT score each morning after waking.
- Breathe through the nose both day and night, including wearing tape over the mouth during sleep (page 64).
- Warm up for 10 minutes by walking and performing a breath hold every minute or so to Simulate High-Altitude Training (page 126).
- Breathe Light to Breathe Right (page 74) during the run by increasing the intensity of exercise while maintaining nasal breathing to develop a reasonably strong air shortage.
- Continue with running and nasal breathing for 20 minutes to 1 hour.
- Midway through the run, practice breath holds to Simulate High-Altitude Training. Exhale and hold the breath for 10 to 40 steps while running at a good pace.
- After the breath hold, resume nasal breathing while simultaneously relaxing the body. Continue to intersperse breath holds every few minutes throughout the run.

- After physical exercise, practice the Breathing Recovery Exercise (page 91).
- Practice one session of Advanced Simulation of High Altitude (page 132) every other day.
- Reduce breathing using the Breathe Light to Breathe Right exercise for 15 minutes last thing at night.
- Fill in the following chart to mark your progress.

BOLT score 30 seconds or more	**Example**	**Day 1**	**Day 2**	**Day 3**	**Day 4**	**Day 5**	**Day 6**	**Day 7**
BOLT	7 A.M. 35 seconds							
Breathe Light Run	10 A.M. 45 minutes							
Simulate High-Altitude Training—Run	Completed during run							
Advanced Simulation of High Altitude	12 P.M. Completed		Day Off		Day Off		Day Off	
Breathe Light before sleep	10:30 P.M. 15 minutes							

Brenda is thirty-two years old, in perfect health, and runs ten miles, four days a week. She is a competitive long-distance runner with a BOLT score of between 35 and 40 seconds. In order to maintain a high BOLT score, physical exercise with a medium to strong air hunger is necessary.

When Brenda feels that she wants to push her body harder and faster, she increases her pace to run as fast as she can with her mouth closed. Often, she is able to maintain nasal breathing while running at her maximum pace. As Brenda's BOLT score is close to 40 seconds, her body can perform high-intensity exercise without needing to revert to breathing through her mouth. Since her breathing is already so efficient, opening the mouth provides no advantage to her performance. Brenda's current training regimen is sufficient to ensure maintenance of her BOLT score.

Oxygen Advantage Program Summary for a BOLT Score of 10 to 30 Seconds Plus

With continued practice each week culminating in an improvement to your BOLT score, more intense exercises can be employed to reach even greater heights. Using the following illustrations, you can check your pathway to success.

Oxygen Advantage Podium

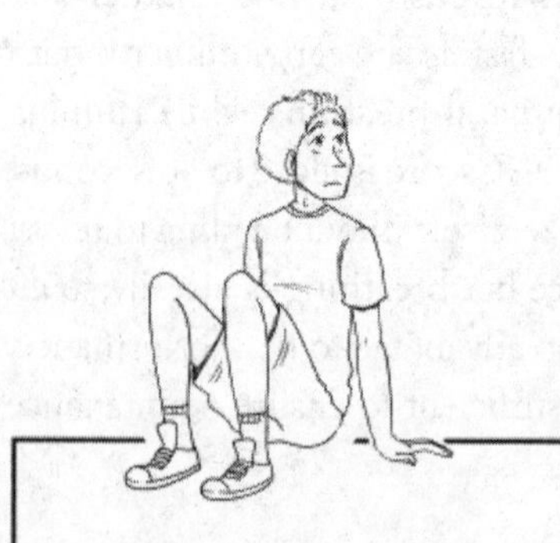

BOLT Score of Less Than 10 Seconds

Measure morning BOLT score

Breathe through the nose both day and night

Practice the Breathing Recovery Exercise (10 minutes by 6 times daily)

Engage in 10 to 15 minutes of slow walking each day with mouth closed

When BOLT score increases to 15 seconds, Breathe Light to Breathe Right (10 minutes by 6 times daily)

BOLT Score of 10 to 20 Seconds

Measure morning BOLT score

Breathe through the nose both day and night

Avoid sighing and taking big breaths

Breathe Light to Breathe Right (10 minutes by 3 times daily)

Breathe Light to Breathe Right—walking or slow jog (30 to 60 minutes daily)

Breathing recovery after physical exercise

BOLT Score of 20 to 30 Seconds and Good Health:

Measure morning BOLT score

Breathe through the nose both day and night

Breathe Light to Breathe Right—sitting (10 minutes by 3 times daily)

Warm up prior to physical exercise

Breathe Light to Breathe Right—fast walk or jog (30–60 minutes per day)

Simulate High-Altitude Training—fast walk or jog

Breathing recovery after physical exercise

BOLT Score of 30 Seconds Plus and Good Health:

Measure morning BOLT score

Breathe through the nose both day and night

Warm up prior to physical exercise

Breathe Light to Breathe Right during physical exercise

Simulate High-Altitude Training—jogging or running

Advanced Simulation of High-Altitude Training every other day

Breathe Light to Breathe Right exercise for 15 minutes before bed

Program for Weight Loss or Obesity (suitable for all BOLT scores)

- Permanently switch to nasal breathing during both day and night.
- Wear tape across the mouth during sleep (page 64).
- Become conscious of your breathing during everyday tasks, allowing it to be calm, relaxed, and quiet.
- Practice Breathe Light to Breathe Right (page 74) for 10 to 15 minutes, 5 times each day. This can be divided into the following:
    - 10 minutes before work
    - 10 minutes during lunch
    - 10 minutes after work
    - 10 minutes (or more) while watching TV in the evening
    - 15 minutes before bed
- Breathe Light to Breathe Right during walking for 30 to 60 minutes per day.
- For those with a BOLT of over 20 seconds, and who are suited to performing breath holds, Simulate High-Altitude Training (page 126) by incorporating 8 to 10

breath holds to achieve a medium air hunger throughout the walk.

- Pay particular attention to hunger sensations, asking yourself whether you really need to eat at that time and stopping when satisfied.
- Fill in the following chart to mark your progress.

Weight Loss	Example	Day 1	Day 2	Day 3	Day 4	Day 5	Day 6	Day 7
BOLT	7:45 A.M. 17 seconds							
Breathe Light	8 A.M. 10 minutes							
Breathe Light	10 minutes							
Breathe Light	12:30 P.M. 10 minutes							
Breathe Light	6 P.M. 10 minutes							
Breathe Light	11:15 P.M. 15 minutes before sleep							
Breathe Light to Breathe Right during walking	3 P.M. Completed							

Earlier on in the book we were introduced to Donna, whose major bugbear was the frustration and feeling of self-defeat when she found her weight increasing again after a successful period of weight loss. To help Donna get her health back on track and to achieve her ideal weight, I asked her to do just three exercises: Breathe Light to Breathe Right, walking with her mouth closed, and practicing breath holds during walking.

Her starting BOLT was 12 seconds, which is typical of someone in a stressful and demanding job with little time devoted to physical exercise. And although Donna did breathe through her nose during the day, she woke up most mornings with a dry mouth, indicating mouth breathing during sleep. Another factor related to her low BOLT score was her difficulty falling asleep. Although she went to bed at a reasonable hour, it often took her two to three hours to finally fall asleep. The result was that she woke up feeling lethargic and unrefreshed, not an ideal scenario when work and family life demanded alertness.

My main goal for Donna was to increase her BOLT score, experience a natural suppression to appetite, and improve her sleep and energy levels.

The most important time for Donna to Breathe Light with a tolerable air hunger was the session directly before she went to sleep. It was not necessary for her to practice this in bed. Instead, Donna usually had some downtime late in the evening when she watched TV, and this was an excellent time for her to reduce her breathing without having to devote special time to the exercise. I suggested that Donna not watch the news or any program involving violence or aggression. This exercise offers an opportunity for the body to relax, not to invoke a stress response.

Because she breathed lightly right before she went to bed, Donna's sleep became deeper, enabling her to wake up 15 minutes earlier to begin her first set for the day. It was also vital that she made a conscious effort to breathe lightly at various times throughout the day. This just involved bringing attention to her breathing, calming it and quieting it so that she felt a light to medium need for air.

Donna found it a challenge to keep her mouth closed during sleep, but within a few nights her body adapted to this new way of breathing. After the fourth night, she found herself sleeping all the way through, requiring less sleep, waking up earlier, and feeling more refreshed.

In addition to practicing her breathing exercises, it was important that Donna paid attention to her sensations of hunger and thirst. Improving the oxygenation of tissues and organs results in the body using food more efficiently, with this leading to a natural suppression of appetite. The advice I gave to Donna was to eat only when hungry and stop when satisfied. By following this simple rule, she automatically had less need for snacking during the day, and even found herself taking lunch at a later time. Eating according to bodily needs is much more important than eating according to a specific time of the day. It is unfortunate that society has dictated when we eat more so than when we actually need the food. Another beneficial side effect that Donna experienced was an increased thirst and demand for water.

Donna experienced a noticeable reduction to her appetite, and her progress with the breathing exercises was reflected in a higher BOLT score. Within two weeks, her BOLT score had increased from 12 to 20 seconds and she had lost six pounds. I encouraged Donna to quicken her progress by incorporating 8 to 10 breath holds into her 20-minute walk to create a medium air hunger. Doing this exercise temporarily lowers oxygen levels to simulate high-altitude training, resulting in a further natural suppression of appetite.

Many people with a low BOLT score, including Donna, avoid physical exercise, as they get breathless far too quickly, making exercise a burdensome chore. As Donna's BOLT score increased to 20 seconds, she felt the need to do more physical exercise. This further helped with her weight normalization, increased energy, and self-esteem.

I told Donna what I tell everyone else: Breathing Light to Breathe Right isn't just an exercise; it's a way of life. How we breathe at all times during the day influences how we feel and our health. When Donna incorporated this into her way of life and maintained a BOLT of 25-plus seconds, her days of yo-yo dieting were over. To her delight, she currently weighs 145 pounds (from a top weight of 175 pounds), and the increased energy and feel-good factor is a great motivator to continue.

Program for Children and Teenagers

- The Nose Unblocking exercise (page 61) is the best exercise for children as it is easy, quick, and measurable.
- Practice the Nose Unblocking Exercise a total of 12 times per day, divided into 2 sets of 6 repetitions. Practice 6 repetitions before breakfast and 6 during the day. The number of paces that a child is able to do should increase by 10 every week, with a goal of reaching a score of 80 to 100 paces.
- While practicing the Nose Unblocking Exercise, I often encourage children to wear paper tape over their mouth. This ensures that the mouth is closed during the exercise, with no air sneaking in (see page 64).
- Wearing the tape while watching TV or going about the house can also be very helpful for a child to get used to breathing through his or her nose only.
- Breathe through the nose throughout the day with the tongue placed in the roof of the mouth. To find out more about mouth breathing and development of the growing face, I suggest that you read my book *Buteyko Meets Dr. Mew: Buteyko Method for Children and Teenagers.*
- Fill in the following chart to mark your progress.

Children and Teenagers	Example	Day 1	Day 2	Day 3	Day 4	Day 5	Day 6	Day 7
Daytime 1st Paces	25							
Daytime 2nd Paces	27							
Daytime 3rd Paces	30							
Daytime 4th Paces	25							
Daytime 5th Paces	28							
Daytime 6th Paces	30							
Evening time 7th Paces	35							
Evening time 8th Paces	35							
Evening time 9th Paces	37							
Evening time 10th Paces	30							
Evening time 11th Paces	40							
Evening time 12th Paces	37							

Marc is seven years old, with nasal obstruction and continuous mouth breathing. Although Marc's physician ruled out asthma, his breathing is audible during rest and eating (much to his parents' consternation). He gets very breathless while playing football, having to take a break from the match to get his breathing back. Furthermore, he snores each night.

Children are great responders to breathing retraining, although a lot of observation is required from parents. I always say that the success of a child's retraining is dependent on the motivation of the parents. At my courses, I motivate parents by discussing the importance of nasal breathing for facial development, concentration, sleep, and overall health. I use my own example of mouth breathing during childhood and how it detrimentally affected my concentration in high school and college. Such an innocuous habit has devastating consequences.

To keep it simple for children and teenagers, I suggest only one exercise and a couple of guidelines to make good progress.

Marc loves football, and the promise of improving his sports performance is the main motivator to put the exercises into practice. When Marc is able to hold his breath for 80 to 100 paces, all he has to do is to practice enough repetitions to maintain this figure. For example, after a few weeks Marc should be able to maintain 80 paces by just doing 3 repetitions of paces each day.

An example of Marc's progress is as follows:

Week 1: 32 paces

Week 2: 37 paces

Week 3: 49 paces

Week 4: 58 paces

Week 5: 70 paces

Week 6: 81 paces

Week 7: 83 paces

Week 8: 79 paces

Week 9: 82 paces

Week 10: 85 paces

Marc's progress on the number of paces that he can hold his breath for will also depend on his everyday breathing. If he does not breathe through his nose, or regularly takes big breaths during the day, then his progress with increasing the number of paces while holding his breath will be a lot slower. Therefore, in addition to practicing the paces, it is vital to ensure nasal and quiet breathing during the day. To help with this, Marc's parents can gently encourage him to breathe quietly any time they hear his breathing. An added incentive for his parents to continue with the exercises is that Marc will have silent breathing during eating, not finding it necessary to gulp large breaths through his mouth while having a mouth full of food.

Appendix
Upper Limits and Safety of Breath Holding

When you hold your breath, you prevent oxygen from entering into your lungs and excess carbon dioxide from been expelled. During a maximum breath hold, the partial pressure of oxygen decreases in the blood, which causes the body to conserve any available oxygen for the heart and brain by constricting the blood vessels that supply nonessential organs. For example, your arms and legs may feel cold as the blood vessels close and the body diverts blood away from them. Another effect is bradycardia, which is the slowing of the heart, causing peripheral blood vessels to constrict, blood pressure to increase, and the spleen to contract, resulting in the "diving response." The diving response is experienced by all air-breathing vertebrates and is an automatic response to a drop in oxygen supply. It is what allows babies and young children to instinctively hold their breath when underwater, and is generally more pronounced in adults who regularly practice breath holding.

During breath holding, the arterial partial pressure of oxygen decreases from its normal level (100 mmHg) while carbon dioxide increases to above its normal level (40 mmHg). Breaking point, or the point at which an individual must release a breath hold, is when oxygen decreases to 62 mmHg and carbon dioxide is 54 mmHg.

While it is extremely difficult for adults to hold their breath to the point of fainting, it has been estimated that consciousness is lost when the oxygen drops to below 27 mmHg and carbon dioxide increases to between 90 and 120 mmHg. The body uses built-in safety mechanisms such as the diving response and fainting to ensure that we do not deprive the brain of oxygen for too long, as doing so can result in brain injury.

The breath-hold exercises described in this book are absolutely safe so long as they are practiced within tolerable limits. However, individuals with high blood pressure, cardiac conditions, type 1 diabetes, or any other serious health concern should not practice holding their breath either during rest or while exercising.

To simulate high-altitude training, it is necessary to hold your breath until you feel a relatively strong air hunger. At the same time, there is no point in overdoing it. It is important to recover your breathing within 2 to 3 breaths. Before practicing breath holding during intense exercise, it is advisable to first achieve a BOLT score of at least 20 seconds. Until then, gentler breath-hold exercises during rest and mild to moderate activity can help you to improve your BOLT score to 20 seconds or more.

Although breath-hold training increases tolerance of carbon dioxide, it is interesting to note that it does not blunt the brain's safety reaction to oxygen deprivation. This is where deliberate breath-hold exercises differ vastly from the physiological condition of sleep apnea, where the breath is held unintentionally during sleep, sometimes leading to severe health problems. If these frightening results are a side effect of breath holding, it may be supposed that intentional breath holding might have the same effects, but studies of elite breath holders have found the results to be quite the opposite. Research by Ivancev and colleagues looked at the breath-holding ability and carbon dioxide sensitivity of breath-hold divers, whose sport potentially puts them at risk of severe oxygen deprivation. With repeated practice, these divers are able to sustain very long breath holds, inducing a severe drop in oxygen without causing brain injury or blackouts. A further study by Joulia and colleagues showed that divers displayed a more pronounced diving response, less of a decrease in oxygen saturation, and greater blood flow.

Stages of Breath Holding

A breath hold can be distinguished by three stages of air shortage ranging from easy to moderate to strong.

In the first stage there is no stimulus by the breathing muscles to resume breathing because carbon dioxide has not yet reached threshold limits. This is known as an easy air shortage.

The second stage is a moderate air shortage. As the breath hold lengthens, carbon dioxide continues to increase in the blood until the concentration reaches your threshold, stimulating the breathing muscles to contract or jerk in an attempt to draw air in. The longer the breath is held, the more frequent are the contractions of the breathing muscles as the body attempts to draw air into the lungs.

The third stage is when the desire to breathe becomes so strong that the individual must resume breathing. This is known as a strong air shortage.

- Easy air shortage: no sensation to breathe.
- Moderate air shortage: from the first involuntary contraction of the breathing muscles until contractions become frequent.
- Strong air shortage: urge to breathe is strong, leading to termination of the breath hold.

Influences on Length of Breath-Hold Time

The three factors that determine the length of breath-hold time are: metabolic rate, tolerance to asphyxia (decreased oxygen levels), and total body gas storage in lungs, blood, and tissues.

Metabolic rate can be reduced through relaxation before and during the breath hold, whereas tolerance to asphyxia is improved by practicing regular breath holds. Other activities that influence breath-hold time are:

- Distraction
- Whether the breath hold follows an inhalation or exhalation
- Whether the athlete hyperventilates before the breath hold

Holding the breath following an inhalation results in a longer breath-hold time since carbon dioxide is diluted in a larger volume of air, meaning the brain's receptors to carbon dioxide are not activated as quickly.

Breath-hold time increases if you take a number of big breaths immediately prior to holding your breath, but this effect is especially dangerous when practiced by swimmers. Taking big breaths immediately before a swim will significantly reduce carbon dioxide in the blood but has little effect on increasing oxygen stores. Because the brain's signal to resume breathing is depleted by this technique, oxygen levels can drop to very low levels before the swimmer feels the need to breathe. This situation can result in the swimmer losing consciousness underwater and in worst-case scenarios can cause death by drowning. The U.S. Navy Seals website warns against this dangerous practice:

> **Important Alert:** It has come to our attention that many men preparing for Special Operations Forces training are practicing breath holding underwater, and there have been several cases of drowning and near drowning recently. **Please do not practice breath holding (under water) without professional supervision.**

Notes

1. The Oxygen Paradox

22 When levels of carbon dioxide: Cheung S. *Advanced Environmental Exercise Physiology.* Champaign, IL: Human Kinetics; 2009.

26 "The carbon dioxide pressure": Bohr C, Hasselbalch K, Krogh A. Concerning a biologically important relationship—the influence of the carbon dioxide content of blood on its oxygen binding. *Skand Arch Physiol* 1904;16:401–12; www.udel.edu/chem/white/C342/Bohr%281904%29.html (accessed August 2012).

26 "an exercising muscle is hot": West JB. *Respiratory Physiology: The Essentials.* Philadelphia: Lippincott Williams & Wilkins; 1995.

27 In general, blood flow: Magarian GJ, Middaugh DA, Linz DH. Hyperventilation syndrome: A diagnosis begging for recognition. *West J Med.* 1983 May; 138(5):733–36.

27 A study by Dr. Daniel M. Gibbs: Gibbs DM. Hyperventilation-induced cerebral ischemia in panic disorder and effects of nimodipine. *Am J Psychiatry.* 1992 Nov;149(11):1589–91.

27 It is well documented that habitual mouth breathing: Kim EJ, Choi JH, Kim KW, et al. The impacts of open-mouth breathing on upper airway space in obstructive sleep apnea: 3-D MDCT analysis. *Eur Arch Otorhinolaryngol.* 2011 Apr;268(4):533–9. Kreivi HR, Virkkula P, Lehto J, Brander P. Frequency of upper airway symptoms before and during continuous positive airway pressure treatment in patients with obstructive sleep apnea syndrome. *Respiration.* 2010;80(6):488–94. Ohki M, Usui N, Kanazawa H, Hara I, Kawano K. Relationship between oral breathing and nasal obstruction in patients with obstructive sleep apnea. *Acta Otolaryngol Suppl.* 1996;523:228–30. Lee SH, Choi JH, Shin C, Lee HM, Kwon SY, Lee SH. How does open-mouth breathing in-

fluence upper airway anatomy? *Laryngoscope.* 2007 Jun;117(6):1102–6. Scharf MB, Cohen AP. Diagnostic and treatment implications of nasal obstruction in snoring and obstructive sleep apnea. *Ann Allergy Asthma Immunol.* 1998 Oct;81(4):279–87; quiz 287–90. Wasilewska J, Kaczmarski M. Obstructive sleep apnea-hypopnea syndrome in children. *Wiad Lek.* 2010;63(3):201–12. Rappai M, Collop N, Kemp S, deShazo R. The nose and sleep-disordered breathing: What we know and what we do not know. *Chest.* 2003 Dec;124(6):2309–23.

28 However, an increase of carbon dioxide opens: A study by Dr. van den Elshout from the department of pulmonary diseases at the University of Nijmegen in the Netherlands explored the effect on airway resistance when there is an increase of carbon dioxide (*hypercapnia*) or a decrease (*hypocapnia*). Altogether, fifteen healthy people and thirty with asthma were involved. The study found that an increase of carbon dioxide resulted in a "significant fall" in airway resistance in both normal and asthmatic subjects. This simply means that the increase of carbon dioxide opened the airways to allow a better oxygen transfer to take place. Interestingly, individuals without asthma also experienced better breathing. Van den Elshout FJ, van Herwaarden CL, Folgering HT. Effects of hypercapnia and hypocapnia on respiratory resistance in normal and asthmatic subjects. *Thorax.* 1991;46(1):28–32.

28 Maintaining normal blood pH: Casiday R, Frey R. *Blood, Sweat, and Buffers: pH Regulation During Exercise. Acid-Base Equilibria Experiment.* www.chemistry.wustl.edu/~edudev/LabTutorials/Buffer/Buffer.html (accessed August 20, 2012).

31 As the late chest physician Claude Lum: Lum LC. Hyperventilation: The tip and the iceberg. *J Psychosom Res.* 1975;19(5–6):375–83.

2. How Fit Are You Really? The Body Oxygen Level Test (BOLT)

32 In fact, studies have shown: A study by Japanese researchers Miharu Miyamura and colleagues from Nagoya University, of ten marathon runners and fourteen untrained individuals found that athletes had a significantly greater tolerance to carbon dioxide at rest when compared with untrained individuals. The study found that for the same amount of exercise, athletes experienced 50 to 60 percent less breathlessness than that of untrained individuals. Miyamura M, Yamashina T, Honda Y. Ventilatory responses to CO_2 rebreathing at rest and during exercise in untrained subjects and athletes. *Jpn J Physiol.* 1976;26(3):245–54.

34 Oxidative stress occurs when: Finaud J, Lac G, Filaire E. Oxidative stress: Relationship with exercise and training. *Sports Med.* 2006;36(4):327–58.

34 It has been said that one: One difference between endurance athletes and nonathletes is decreased ventilatory responsiveness to hypoxia (low oxygen) and

hypercapnia (higher carbon dioxide). Scoggin CH, Doekel RD, Kryger MH, Zwillich CW, Weil JV. Familial aspects of decreased hypoxic drive in endurance athletes. *J Appl Physiol.* 1978 Mar;44(3):464–8.

In a paper entitled "Low exercise ventilation in endurance athletes" that was published in *Medicine and Science in Sports,* the authors found that nonathletes breathe far heavier and faster to changes in oxygen and carbon dioxide when compared with endurance athletes at equal workloads. The authors observed that the lighter breathing of the athlete group may explain the link between "low ventilatory chemosensitivity and outstanding endurance athletic performance." Martin BJ, Sparks KE, Zwillich CW, Weil JV. Low exercise ventilation in endurance athletes. *Med Sci Sports.* 1979 Summer;11(2):181–5.

35 Studies have shown that athletic ability: In a study published in the *Journal of Applied Physiology* that compared thirteen athletes and ten nonathletes, the athletes' response to increased carbon dioxide was 47 percent of that recorded by the nonathlete controls. The authors noted that athletic ability to perform during lower oxygen pressure and higher carbon dioxide pressure corresponded to maximal oxygen uptake or VO_2 max. Byrne-Quinn E, Weil JV, Sodal IE, Filley GF, Grover RF. Ventilatory control in the athlete. *J Appl Physiol.* 1971 Jan;30(1):91–8. In another study conducted at the Research Centre of Health, Physical Fitness and Sports at Nagoya University in Japan, researchers evaluated nine initially untrained college students. Five out of the nine students took up physical training for three hours a day, three times a week for four years. The researchers found that VO_2 max increased after training and the response of breathing to increased arterial carbon dioxide decreased significantly during each training period. Moreover, CO_2 responsiveness was found to correlate negatively with maximum oxygen uptake in four out of the five trained subjects. Similarly to the previous study, subjects with reduced sensitivity to CO_2 experienced increased delivery of oxygen to working muscles. Miyamura M, Hiruta S, Sakurai S, Ishida K, Saito M. Effects of prolonged physical training on ventilatory response to hypercapnia. *Tohoku J Exp Med.* 1988 Dec;156 Suppl:125–35.

35 There is a strong association: Saunders PU, Pyne DB, Telford RD, Hawley JA. Factors affecting running economy in trained distance runners. *Sports Med.* 2004;34(7):465–85.

36 Researchers investigating reduced breathing found that running economy: Scientists investigated whether controlling the number of breaths during swimming could improve both swimming performance and running economy. A paper published in the *Scandinavian Journal of Medicine and Science in Sports* involved eighteen swimmers, ten men and eight women, who were assigned to

two groups. The first group was required to take only 2 breaths per length and the second group 7 breaths. As swimming is one of the few sports that naturally limits breath intake, it is often of interest to scientists since reducing the amount of air consumed during training adds an additional challenge to the body and may lead to improvements in respiratory muscle strength. Interestingly, the researchers found that running economy improved by 6 percent in the group that performed reduced breathing during swimming. Lavin KM, Guenette JA, Smoliga JM, Zavorsky GS. Controlled-frequency breath swimming improves swimming performance and running economy. *Scand J Med Sci Sports.* 2015 Feb;25(1):16–24.

37 As far back as 1975: Stanley NN, Cunningham EL, Altose MD, Kelsen SG, Levinson RS, Cherniack NS. Evaluation of breath holding in hypercapnia as a simple clinical test of respiratory chemosensitivity. *Thorax.* 1975 Jun;30 (3):337–43.

Japanese researcher Nishino acknowledged breath holding as one of the most powerful methods to induce the sensation of breathlessness, and that the breath hold test "gives us much information on the onset and endurance of dyspnea (breathlessness)." The paper noted two different breath-hold tests as providing useful feedback on breathlessness. According to Nishino, because holding of the breath until the first definite desire to breathe is not influenced by training effect or behavioral characteristics, it can be deduced to be a more objective measurement of breathlessness. Nishino T. Pathophysiology of dyspnea evaluated by breath-holding test: Studies of furosemide treatment. *Respir Physiol Neurobiol.* 2009 May 30;167(1):20–5.

39 Since carbon dioxide is the primary: Stanley et al. 1975, 337–43.

40 "If a person breath holds after a normal exhalation": McArdle W, Katch F, Katch V. *Exercise Physiology: Energy, Nutrition, and Human Performance.* 7th ed. Philadelphia: Lippincott Williams & Wilkins; 2010:289.

40 Breath-hold measurements have also been used: The department of physiotherapy at the University of Szeged, Hungary, conducted a study that investigated the relationship between breath-hold time and physical performance in patients with cystic fibrosis. Eighteen patients with varying stages of cystic fibrosis were studied to determine the value of the breath-hold time as an index of exercise tolerance. The breath-hold times of all patients were measured. Oxygen uptake (VO_2) and carbon dioxide elimination were measured breath by breath as the patients exercised. The researchers found a significant correlation between breath-hold time and VO_2 (oxygen uptake), concluding "that the voluntary breath-hold time might be a useful index for prediction of the exercise tolerance of CF patients." Taking this one step further, increasing the BOLT of patients with CF corresponds to greater oxygen uptake and reduced breath-

lessness during physical exercise. Barnai M, Laki I, Gyurkovits K, Angyan L, Horvath G. Relationship between breath-hold time and physical performance in patients with cystic fibrosis. *Eur J Appl Physiol.* 2005 Oct;95(2–3):172–8. Results from a study of thirteen patients with acute asthma concluded that the magnitude of breathlessness, breathing frequency, and breath-hold time was correlated with severity of airflow obstruction and, secondly, that breath-hold time varies inversely with the magnitude of breathlessness when it is present at rest. In simple terms, the lower the breath-hold time of asthmatics, the greater the breathing volume and breathlessness. Pérez-Padilla R, Cervantes D, Chapela R, Selman M. Rating of breathlessness at rest during acute asthma: Correlation with spirometry and usefulness of breath-holding time. *Rev Invest Clin.* 1989 Jul–Sep;41(3):209–13.

3. Noses Are for Breathing, Mouths Are for Eating

52 Mouth breathing activates the upper chest: Swift AC, Campbell IT, McKown TM. Oronasal obstruction, lung volumes, and arterial oxygenation. *Lancet.* 1988 Jan;1(8577):73–75.

52 Dentists and orthodontists have also: Harari D, Redlich M, Miri S, Hamud T, Gross M. The effect of mouth breathing versus nasal breathing on dentofacial and craniofacial development in orthodontic patients. *Laryngoscope.* 2010 Oct;120(10):2089–93. D'Ascanio L, Lancione C, Pompa G, Rebuffini E, Mansi N, Manzini M. Craniofacial growth in children with nasal septum deviation: A cephalometric comparative study. *Int J Pediatr Otorhinolaryngol.* 2010 Oct;74(10):1180–83. Baumann I, Plinkert PK. Effect of breathing mode and nose ventilation on growth of the facial bones. *HNO.* 1996 May;44(5):229–34. Tourne LP. The long face syndrome and impairment of the nasopharyngeal airway. *Angle Orthod.* 1990 Fall;60(3):167–76.

52 One of his observations: Price W (ed.). *Nutrition and Physical Degeneration.* 8th ed. La Mesa, CA: Price-Pottenger Nutrition Foundation; 2008:55.

53 In comparison, Catlin: Catlin G (ed.). *Letters and Notes on the Manners, Customs, and Condition of the North American Indians.* New York: Wiley & Putnam; 1842.

53 Most high-performance cars cannot: Sutcliffe S. Bugatti Veyron online review (2005). www.autocar.co.uk/car-review/bugatti/veyron/first-drives/bugatti-veyron (accessed September 2, 2014).

54 With such incredible efficiency: Burton M, Burton R. (eds.). *The International Wildlife Encyclopedia.* 3rd ed. New York: Marshall Cavendish Corp.; 2002:403.

54 The same is true: Morgan E. Aquatic Ape Theory. Primitivism. www.primitivism.com/aquatic-ape.htm (accessed September 2, 2014).

54 Charles Darwin was: Ibid.

54 Birds, for example: Pelecaniformes. Wikipedia. en.wikipedia.org/wiki/Pelecaniformes (accessed September 2, 2014).

54 Guinea pigs and rabbits: Nixon JM. Breathing pattern in the guinea-pig. *Lab Anim.* 1974;8:71–7. Hernandez-Divers SJ. The rabbit respiratory system: Anatomy, physiology, and pathology. Proceedings of the Association of Exotic Mammal Veterinarians Scientific Program. Providence, RI; 2007:61–8.

55 Experience tells the: Jackson PGG, Cockcroft PD (eds.). *Clinical Examination of Farm Animals.* Oxford, UK, and Malden MA: Blackwell Science; 2002:70.

55 The late Dr. Maurice Cottle: Timmons BH, Ley R (eds.). *Behavioral and Psychological Approaches to Breathing Disorders.* New York: Springer; 1994.

56 In the yoga: Ramacharaka Y (ed.). Nostril vs. mouth breathing. In: *The Hindu-Yogi Science of Breath.* Waiheke Island, New Zealand: Floating Press; 1903.

56 Nose breathing imposes: Timmons, Ley (eds.), *Behavioral and Psychological Approaches.*

56 Nasal breathing warms: Fried R (ed.). *Hyperventilation Syndrome: Research and Clinical Treatment* (Johns Hopkins Series in Contemporary Medicine and Public Health). Baltimore, MD: Johns Hopkins University Press; 1987.

57 Nasal breathing removes: Ibid.

57 Nasal breathing during physical: Morton AR, King K, Papalia S, Goodman C, Turley KR, Wilmore JH. Comparison of maximal oxygen consumption with oral and nasal breathing. *Aust J Sci Med Sport.* 1995 Sep;27(3):51–5.

57 As discussed in the next section: Vural C, Güngör A. Nitric oxide and the upper airways: Recent discoveries. *Tidsskr Nor Laegeforen.* 1999 Nov 10;119(27): 4070–2. Doctors Maria Belvisi and Peter Barnes and colleagues from the National Heart and Lung Institute in the United Kingdom demonstrated that one of the roles of nitric oxide includes dilation of the smooth muscles surrounding the airways. Belvisi MG, Stretton CD, Yacoub M, Barnes PJ. Nitric oxide is the endogenous neurotransmitter of bronchodilator nerves in humans. *Eur J Pharmacol.* 1992 Jan 14;210(2):221–2. Djupesland PG, Chatkin JM, Qian W, Haight JS. Nitric oxide in the nasal airway: A new dimension in otorhinolaryngology. *Am J Otolaryngol.* 2001 Jan–Feb; 22(1):19–32. Lundberg JO. Nitric oxide and the paranasal sinuses. *Anat Rec (Hoboken).* 2008 Nov;291(11):1479–84. Vural C, Güngör A. Nitric oxide and the upper airways: Recent discoveries. *Kulak Burun Bogaz Ihtis Derg.* 2003 Jan;10(1):39–44.

57 Mouth-breathing children are at greater risk: Okuro RT, Morcillo AM, Ribeiro MÂ, Sakano E, Conti PB, Ribeiro JD. Mouth breathing and forward head posture: Effects on respiratory biomechanics and exercise capacity in children. *J Bras Pneumol.* 2011 Jul–Aug;37(4):471–9. Conti PB, Sakano E, Ribeiro MA,

Schivinski CI, Ribeiro JD. Assessment of the body posture of mouth-breathing children and adolescents. *J Pediatr (Rio J).* 2011 Jul–Aug;87(4):357–63.

57 A dry mouth also increases acidification: Orthodontists online community. Mouth Breathing. orthofree.com/fr/default.asp?contentID=2401 (accessed January 7, 2015).

57 Mouth breathing causes: Ibid.

57 Breathing through the mouth: Kim et al. 2010, 533–9. Kreivi et al. 2010, 488–94. Ohki et al. 1996, 228–30. Lee et al. 2007 Jun, 1102–6. Scharf, Cohen 1998 Oct, 279–87; quiz 287–90.

58 When the first article appeared discussing: Chang, HR (ed.). *Nitric Oxide, the Mighty Molecule: Its Benefits for Your Health and Well-Being.* Jacksonville, FL: Mind Society; 2011.

58 And although nitric oxide: Ibid.

58 In 1992, nitric oxide: Culotta E, Koshland DE Jr. NO news is good news. *Science.* 1992 Dec 18;258(5090):1862–5.

58 In 1998, Robert F. Furchgott: Raju, TN. The Nobel chronicles. 1998: Robert Francis Furchgott (b 1911), Louis J Ignarro (b 1941), and Ferid Murad (b 1936). *Lancet.* 2000 Jul 22;356(9226):346. Rabelink, AJ. Nobel prize in medicine and physiology 1998 for the discovery of the role of nitric oxide as a signalling molecule. *Ned Tijdschr Geneeskd.* 1998 Dec 26;142(52):2828–30.

58 When I first began: Ignarro L. *NO More Heart Disease: How Nitric Oxide Can Prevent—Even Reverse—Heart Disease and Strokes.* Rprt. New York: St. Martin's Press; 2006. Cartledge J, Minhas S, Eardley I. The role of nitric oxide in penile erection. *Expert Opin Pharmacother.* 2001 Jan;2(1):95–107. Toda N, Ayajiki K, Okamura T. Nitric oxide and penile erectile function. *Pharmacol Ther.* 2005 May;106(2):233–66. Chang, *Nitric Oxide, the Mighty Molecule*; 2012.

58 Nitric oxide is produced: Lundberg JO, Weitzberg E. Nasal nitric oxide in man. *Thorax.* 1999;(54):947–52. Chang, *Nitric Oxide, the Mighty Molecule;* 2012. Lundberg JO. Airborne nitric oxide: Inflammatory marker and aerocrine messenger in man. *Acta Physiol Scand Suppl.* 1996;633:1–27.

58 Scientific findings have shown: Maniscalco M, Sofia M, Pelaia G. Nitric oxide in upper airways inflammatory diseases. *Inflamm Res.* 2007 Feb;56(2):58–69. Lundberg 1996, 1–27. Lundberg, Weitzberg 1999, 947–52.

59 "During inspiration through the nose": Lundberg, Weitzberg 1999, 947–52.

59 This short-lived gas: Roizen MF, Oz MC. *You on a Diet: The Owner's Manual for Waist Management.* Rev. ed. New York: Free Press; 2009.

59 It helps to prevent high blood pressure: Ignarro, *NO More Heart Disease;* 2006.

59 The potency of nitric oxide: Cartledge, Minhas, Eardley 2001, 95–107. Toda, Ayajiki, Okamura 2005 May, 233–66.

60 In a study of a group of thirty-three: Gunhan K, Zeren F, Uz U, Gumus B, Unlu H. Impact of nasal polyposis on erectile dysfunction. *Am J Rhinol Allergy.* 2011 Mar–Apr;25(2):112–5.

60 And women can benefit: Roizen, Oz, *You on a Diet,* 2009.

60 In addition to improving your sex: Chang, *Nitric Oxide, the Mighty Molecule*; 2012.

60 Most important for athletes: Fried (ed.), *Hyperventilation Syndrome,* 1987.

60 They concluded that humming: Weitzberg E, Lundberg JO. Humming greatly increases nasal nitric oxide. *Am J Respir Crit Care Med.* 2003 Jul 15;166(2):144–5.

61 The results were an amazing: Three months following the instruction, results as published in the leading European rhinitis journal *Clinical Otolaryngology* showed a 70 percent reduction of symptoms such as nasal stuffiness, poor sense of smell, snoring, trouble breathing through the nose, trouble sleeping, and having to breathe through the mouth. Adelola OA, Oosthuiven JC, Fenton JE. Role of Buteyko breathing technique in asthmatics with nasal symptoms. *Clin Otolaryngol.* 2013 Apr;38(2):190–1.

4. Breathe Light to Breathe Right

68 "And the third level": Pei C. *Qi Gong for Beginners.* DVD. Body Wisdom; 2009.

68 The traditional Chinese philosophy: Blofeld J. *Taoism: The Road to Immortality.* Boulder, CO: Shambhala; 1978.

68 While the kid's manners: *Lavell Crawford kids on fat people.* www.youtube.com/all_comments?v=U6rFzngemUE (accessed September 2, 2014).

68 Authentic professional yoga: Researcher Miharu Miyamura investigated the sensitivity to carbon dioxide during respiration of 1 breath per minute for an hour by a professional Hatha yogi. Results showed that authentic yoga practitioners have reduced sensitivity to carbon dioxide. Miyamura M, Nishimura K, Ishida K, Katayama K, Shimaoka M, Hiruta S. Is man able to breathe once a minute for an hour? The effect of yoga respiration on blood gases. *Jpn J Physiol.* 2002 Jun;52(3):313–6.

5. Secrets of Ancient Tribes

80 Tom theorized that: Tom Piszkin. Personal e-mail to Patrick McKeown, August 2014.

80 His results showed: Babbitt B. Gun shot at the Oakland Coliseum. *Competitor Magazine.* 1988. www.ttinet.com/tf/about2.htm (accessed July 1, 2012).

80 When researchers studied: Douillard J. *Perfect Health for Kids: Ten Ayurvedic Health Secrets Every Parent Must Know.* Berkeley, CA: North Atlantic Books; 2004.

81 To date, he: Sebring L. What does it really feel like to be a healthy human? *Whole Human* (blog). the-whole-human.com/article/lane-sebring-md/what-does-it-really-feel-be-healthy-human (accessed June 10, 2013).

81 Just like our ancestors: Ibid.

84 Incorporating this concept: Woorons X, Mollard P, Pichon A, Duvallet A, Richalet JP, Lamberto C. Effects of a 4-week training with voluntary hypoventilation carried out at low pulmonary volumes. *Respir Physiol Neurobiol.* 2008 Feb 1;160(2):123–30.

86 The vast majority of sports: LaBella CR, Huxford MR, Grissom J, Kim KY, Peng J, Christoffel KK. Effect of neuromuscular warm-up on injuries in female soccer and basketball athletes in urban public high schools: Cluster randomized controlled trial. *Arch Pediatr Adolesc Med.* 2011 Nov;165(11):1033–40. Woods K, Bishop P, Jones E. Warm-up and stretching in the prevention of muscular injury. *Sports Med.* 2007;37(12):1089–99. Shellock FG, Prentice WE. Warming-up and stretching for improved physical performance and prevention of sports-related injuries. *Sports Med.* 1985 Jul–Aug;2(4):267–78.

92 Having successfully completed forty: Danny Dreyer, Founder & President. ChiRunning. www.chirunning.com/about/staff-profile/danny-dreyer (accessed September 2, 2014).

92 A firm exponent of nasal breathing: Dreyer D, Dreyer K. *ChiRunning: A Revolutionary Approach to Effortless, Injury-Free Running.* Rev. ed. New York: Simon & Schuster; 2009:54.

6. Gaining the Edge—Naturally

95 According to the United States: Wilber RL. Application of altitude/hypoxic training by elite athletes. *Med Sci Sports Exerc.* 2007 Sep;39(9):1610–24.

96 This addition of blood: Ekblom BT. Blood boosting and sport. *Baillieres Best Pract Res Clin Endocrinol Metab.* 2000 Mar;14(1):89–98.

96 By the early 1990s: Sawka MN, Joyner MJ, Miles DS, Robertson RJ, Spriet LL, Young AJ. American College of Sports Medicine position stand: The use of blood doping as an ergogenic aid. *Med Sci Sports Exerc.* 1996 Jun;28(6):i–viii.

97 Early reports involved: Walsh D. *From Lance to Landis: Inside the American Doping Controversy at the Tour de France.* New York: Ballantine Books; 2007.

97 During the race: Fotheringham W. *Put Me Back on My Bike: In Search of Tom Simpson.* New ed. London: Yellow Jersey Press; 2007. The death of Tom Simpson. BBC World Service. www.bbc.co.uk/programmes/p00hts7t (accessed September 2, 2014).

97 With a determined effort: Remembering a sensation. BBC. www.bbc.co.uk/insideout/northeast/series6/cycling.shtml (accessed September 2, 2014).

97 Later, investigators would: Ibid.

97 "To Lance's way": Ungoed-Thomas J. Lance Armstrong "given drugs in lunch bag," claims teammate Tyler Hamilton. *Sunday Times.* September 2, 2012.

98 Summed up in the statement: USADA. Statement from USADA CEO Travis T. Tygart Regarding the U.S. Postal Service Pro Cycling Team Doping Conspiracy. cyclinginvestigation.usada.org (accessed January 14, 2015).

98 When Winfrey asked: Oprah Winfrey. Interview with Lance Armstrong. www.oprah.com/own/Lance-Armstrong-Confesses-to-Oprah-Video (accessed September 2, 2014).

98 Looking back, Swart: Pegden E. Swart vindicated by Armstrong report. *Waikato Times.* October 12, 2012. Available at www.stuff.co.nz/sport/other-sports/7805732/Swart-vindicated-by-Armstrong-report (accessed January 14, 2015).

99 "I've always understood": *Rough Rider.* RTE Television. www.rte.ie/tv/programmes/roughrider.html (accessed September 2, 2014).

100 Levels of hemoglobin: Hemoglobin. MedlinePlus. www.nlm.nih.gov/medlineplus/ency/article/003645.htm (accessed August 15, 2012).

100 Hematocrit is usually: Hematocrit. MedlinePlus. www.nlm.nih.gov/medlineplus/ency/article/003646.htm (accessed April 20, 2013).

101 Athletes still use: Levine BD, Stray-Gundersen J. A practical approach to altitude training: Where to live and train for optimal performance enhancement. *Int J Sports Med.* 1992 Oct;13 Suppl 1:209–12.

101 To limit the detraining: Levine BD. Intermittent hypoxic training: Fact and fancy. *High Alt Med Biol.* 2002 Summer;3(2):177–193. Levine BD. Should "artificial" high altitude environments be considered doping? *Scand J Med Sci Sports.* 2006 Oct;16(5):297–301. Levine BD, Stray-Gundersen J. "Living high-training low": Effect of moderate-altitude acclimatization with low-altitude training on performance. *J Appl Physiol.* 1997 Jul;83(1):102–12.

102 Levine and Stray-Gundersen: Levine, Stray-Gundersen 1997, 102–12.

102 These improvements were: Ibid.

102 Furthermore, the increase: Stray-Gundersen J, Chapman RF, Levine BD. "Living high–training low" altitude training improves sea level performance in male and female elite runners. *J Appl Physiol.* 2001 Sep;91(3):1113–20.

102 During the 2006 Torino: Wallechinsky D. *The Complete Book of the Winter Olympics.* Turin 2006 ed. Wilmington, DE: Sport Media Publishing; 2005.

103 The authors of the study concluded that although: Moderate-intensity aerobic training that improves the maximal aerobic power does not change anaerobic capacity, and adequate high-intensity intermittent training may improve both anaerobic and aerobic energy supplying systems significantly. Tabata I, Nishimura K, Kouzaki M, et al. Effects of moderate-intensity endurance and

high-intensity intermittent training on anaerobic capacity and VO_2 max. *Med Sci Sports Exerc.* 1996 Oct;28(10):1327–30.

103 Posttrial results: Bailey SJ, Wilkerson DP, Dimenna FJ, Jones AM. Influence of repeated sprint training on pulmonary O_2 uptake and muscle deoxygenation kinetics in humans. *J Appl Physiol.* 2009 Jun;106(6):1875–87.

103 This means that the athletes: Jones A. Understand the body's use of oxygen during exercise: Oxygen kinetics—start smart for a mean finish! *Sports Performance Bulletin.* www.pponline.co.uk/encyc/understand-the-bodys-use-of-oxygen-during-exercise-36326 (accessed April 20, 2013). Hagberg JM, Hickson RC, Ehsani AA, Holloszy JO. Faster adjustment to and recovery from submaximal exercise in the trained state. *J Appl Physiol Respir Environ Exerc Physiol.* 1980 Feb;48(2):218–24.

105 For hundreds of thousands of years: Rahn H, Yokoyama T. *Physiology of Breath-Hold Diving and the Ama of Japan.* Washington, D.C.: National Academy of Sciences–National Research Council; 1965:369.

105 and some evolutionary theorists: Hardy A. Was man more aquatic in the past? *New Scientist.* March 17, 1960. Hardy A. Was there a *Homo aquaticus*? *Zenith.* 1977;15(1): 4–6.

106 Generally, most humans: World records. Association Internationale pour le Développement de l'Apnée. www.aidainternational.org/competitive/worlds-records (accessed July 6, 2012).

106 A number of studies have sought: Isbister JP. Physiology and pathophysiology of blood volume regulation. *Transfus Sci.* 1997 Sep;18(3):409–23. Koga T. Correlation between sectional area of the spleen by ultrasonic tomography and actual volume of the removed spleen. *J Clin Ultrasound.* 1979 Apr;7(2):119–20. Erika Schagatay is the director of research at Mid Sweden University. Her interest in physiology began after she met native breath-hold divers from several tribes, including Japanese *ama* and Indonesian Suku Laut and Bajau, who were able to hold their breath for far longer than medical literature stated was possible. Schagatay has completed a number of studies on the effects of holding the breath on both trained and untrained breath-hold divers. People. Mid Sweden University. www.miun.se/en/Research/Our-Research/Research-groups/epg/About-EPG/People (accessed August 29, 2012).

One of Schagatay's studies involved twenty healthy volunteers, including ten who had their spleens removed, to determine the adaptations caused by short-term breath holding. The volunteers performed 5 breath holds of maximum duration (as long as possible for each individual) with a 2-minute rest in between each. The results found that the volunteers with spleens showed a 6.4 percent increase in hematocrit (Hct) and a 3.3 percent increase in hemoglobin concentration (Hb) following the breath holds. This means that after just 5 breath holds, the oxygen-carrying capacity of the blood was significantly improved. However, for the individuals who had their spleens removed, there were no recorded changes to the blood resulting from breath holding. Schagatay E, Andersson JP, Hallén M, Pålsson B. Selected contribution: Role of spleen emptying in prolonging apneas in humans. *Journal of Applied Physiology.* 2001 Apr;90(4):1623–9.

During a separate study by Schagatay, seven male volunteers performed 2 sets of 5 breath holds to near maximal duration, one in air and the other with their faces immersed in water. Each breath hold was separated by 2 minutes of rest and each set separated by 20 minutes. Both Hct and Hb concentration increased by approximately 4 percent across both series of breath holds—in air and in water. Schagatay E, Andersson JP, Nielsen B. Hematological response and diving response during apnea and apnea with face immersion. *Eur J Appl Physiol.* 2007 Sep;101(1):125–32.

106 The spleen is an organ: Isbister 1997, 409–23.

106 This means that after as few: Schagatay, Andersson, Nielsen 2007 Sep, 125–32.

106 breath-hold divers peaked: A study by Baković et al. from University of Split School of Medicine, Croatia, was conducted to investigate spleen responses resulting from 5 maximal breath holds. Ten trained breath-hold divers, ten untrained volunteers, and seven volunteers who had their spleen removed were recruited. The subjects performed 5 maximum breath holds with their face immersed in cold water, and each breath hold was separated by a 2-minute rest. The duration of the breath holds peaked at the third attempt, with breath-hold divers reaching 143 seconds, untrained divers reaching 127 seconds, and splenectomized persons achieving 74 seconds. Spleen size decreased by a total of 20 percent in both breath-hold divers and the untrained volunteers. Researchers concluded that the results show rapid, probably active contraction of the spleen in response to breath hold in humans. Rapid spleen contraction and its slow recovery may contribute to prolongation of successive, briefly repeated breath-hold attempts. Baković D, Valic Z, Eterović D, et al. Spleen volume and blood flow response to repeated breath-hold apneas. *J Appl Physiol.* 2003 Oct;95(4):1460–6.

106 Not only that but: Ibid.

107 While these studies generally: In a paper by Dr. Espersen and colleagues from Herlev Hospital, University of Copenhagen, Denmark, splenic contraction was found to take place even with very short breath holds of 30 seconds. However, the strongest contraction of the spleen was as it released blood cells into circulation, occurring when a subject held their breath for as long as possible. Espersen

K, Frandsen H, Lorentzen T, Kanstrup IL, Christensen NJ. The human spleen as an erythrocyte reservoir in diving-related interventions. *J Appl Physiol.* 2002 May;92(5):2071–9.

107 However, the strongest contractions: Ibid.

107 Another useful piece of: This study in particular provides pertinent information about the consequence of breath holding: Since there was no visible increase in the results of breath holding with the subjects' faces immersed in water, the authors concluded that the breath hold, or its consequences, is the major stimulus evoking splenic contraction. Schagatay, Andersson, Nielsen 2007 Sep, 125–32.

107 Performing just 3 to 5 breaths: Ibid.

108 While this reduces the: In his doctoral thesis entitled "Haematological changes arising from spleen contraction during breath hold and altitude in humans," Matt Richardson investigated the role played by higher levels of carbon dioxide.

Eight non-divers performed 3 sets of breath holds on three separate days under different starting conditions, varying the levels of carbon dioxide available to the subjects before each test. The first test was preceded by the breathing of 5 percent CO_2 in oxygen (hypercapnic), the second with pre-breathing of 100 percent oxygen (normocapnic), and the third with hyperventilation of 100 percent oxygen (hypocapnic).

The duration of each breath hold was kept constant in all 3 sets, and baseline values of Hb and Hct were the same for all conditions. After the 3 breath holds, the increase in Hb in the hypercapnic (higher carbon dioxide) trial was 9.1 percent greater than in the normal carbon dioxide trial (normocapnic) and 71.1 percent greater than in the lower carbon dioxide trial (hypocapnic). Richardson concluded that an increased capnic stimulus during breath hold may elicit a stronger spleen response and subsequent Hb increase than breath hold preceded by hyperventilation. Richardson, MX. Hematological changes arising from spleen contraction during apnea and altitude in humans. Doctoral dissertation. Mid Sweden University; 2008.

108 Higher levels of carbon dioxide: Ibid.

108 By exhaling and holding the breath: Dillon WC, Hampl V, Shultz PJ, Rubins JB, Archer SL. Origins of breath nitric oxide in humans. *Chest.* 1996 Oct;110(4):930–8.

109 One of the functions of EPO: Joyner MJ. VO_2MAX, blood doping, and erythropoietin. *Br J Sports Med.* 2003 Jun;37(3):190–191. Lemaître F, Joulia F, Chollet D. Apnea: A new training method in sport? *Med Hypotheses.* 2010 Mar;74(3):413–5.

109 Breath holding is an effective: Lemaître, Joulia, Chollet 2010 Mar, 413–5.

109 The concentration of EPO: De Bruijn and colleagues from the department of natural sciences, Mid Sweden University, investigated whether subjecting the body to lower oxygen levels by holding the breath could increase EPO concentration. The study involved ten healthy volunteers performing 3 sets of 5 maximum duration breath holds, with each set separated by 10 minutes of rest. Results showed that EPO concentration increased by 24 percent, peaking three hours after the final breath hold and returning to baseline two hours later. De Bruijn R, Richardson M, Schagatay E. Increased erythropoietin concentration after repeated apneas in humans. *Eur J Appl Physiol.* 2008 Mar;102(5):609–13.

109 A clear example: Cahan C, Decker MJ, Arnold JL, Goldwasser E, Strohl KP. Erythropoietin levels with treatment of obstructive sleep apnea. *J Appl Physiol.* 1995 Oct;79(4):1278–85. A study by Winnicki and colleagues from the Medical University of Gdansk, Poland, tested the hypothesis that the repetitive lowering of oxygen levels from breath holds during sleep apnea increase EPO. The study involved eighteen severe and ten very mild patients. Results showed a 20 percent increase to EPO in patients with severe obstructive sleep apnea, which decreased following elimination of the breath holds by treatment. Winnicki M, Shamsuzzaman A, Lanfranchi P, et al. Erythropoietin and obstructive sleep apnea. *Am J Hypertens.* 2004 Sep;17(9):783–6.

111 As U.S. Army general: Patton GS Jr. *Third Army, Standard Operating Procedures, 1944.* historicaltextarchive.com/sections.php?action=read&artid=384 (accessed September 2, 2014).

112 Studies with athletes have demonstrated: Lemaître F, Polin D, Joulia F, et al. Physiological responses to repeated apneas in underwater hockey players and controls. *Undersea Hyperb Med.* 2007 Nov–Dec;34(6):407–14. Woorons X, Bourdillon N, Vandewalle H, et al. Exercise with hypoventilation induces lower muscle oxygenation and higher blood lactate concentration: Role of hypoxia and hypercapnia. *Eur J Appl Physiol.* 2010 Sep;110(2):367–77.

114 Dr. Joseph Mercola: Mercola J. Baking soda uses: To remove splinters—and to address many other health needs. Mercola.com. August 27, 2012. articles.mercola.com/sites/articles/archive/2012/08/27/baking-soda-natural-remedy.aspx (accessed June 10, 2013).

114 The therapeutic potential: Marty Pagel, PhD, awarded $2 million NIH grant to study impact of baking soda on breast cancer. University of Arizona Cancer Center. March 21, 2012. azcc.arizona.edu/node/4187 (accessed August 10, 2012).

114 Over the years many: J. Edge and colleagues at the University of Australia in Perth conducted a study of the effects of bicarbonate of soda on the ability of muscles to neutralize the acid that accumulates during high-intensity train-

ing. In Edge's study, sixteen recreationally active women were recruited and randomly placed in two groups of eight. One group ingested bicarbonate of soda and the other ingested a placebo. The results showed that the bicarbonate group experienced greater improvements in lactate threshold and time to fatigue. Their working muscles were better able to neutralize the acid resulting from training, showing improvements to endurance performance. Edge J, Bishop D, Goodman C. Effects of chronic NaHCO3 ingestion during interval training on changes to muscle buffer capacity, metabolism, and short-term endurance performance. *J Appl Physiol.* 2006 Sep;101(3):918–25.

In a study at the University of Bedfordshire in the UK, researchers investigated the effects of sodium bicarbonate on maximum breath-hold time. Eight recreational breath-hold divers were recruited to partake in two bouts of 3 monitored breath holds while their faces were immersed in water. Following the study, the authors suggested that ingestion of bicarbonate of soda before breath holds prolongs maximum breath-hold time by approximately 8.6 percent. Sheard PW, Haughey H. Sodium bicarbonate and breath-hold times. Effects of sodium bicarbonate on voluntary face immersion breath-hold times. *Undersea Hyperb Med.* 2007 Mar–Apr;34(2):91–7.

Researchers from the Academy of Physical Education in Katowice, Poland conducted a study to evaluate the effects of oral administration of sodium bicarbonate on swim performance in competitive youth swimmers. The swimmers completed two time trials: one after ingestion of bicarbonate and one after ingestion of a placebo. Total time for the 4 x 50m test trial improved from 1.54.28 to 1.52.85 s. In addition, bicarbonate had a significant effect on resting blood pH. Researchers concluded that the ingestion of sodium bicarbonate in youth athletes is an effective buffer during high-intensity interval swimming and suggested that such a procedure may be used in youth athletes to increase training intensity and swimming performance in competition at distances from 50 to 200m. Zajac A, Cholewa J, Poprzecki S, Waskiewicz Z, Langfort J. Effects of sodium bicarbonate ingestion on swim performance in youth athletes. *J Sports Sci Med.* 2009 Mar 1;8(1):45–50.

114 During high-intensity: Edge, Bishop, Goodman 2006 Sep, 918–25.

114 By ingesting bicarbonate: Ibid.

114 The ingestion of bicarbonate: Sheard, Haughey. 2007 Mar–Apr, 91–7.

114 Researchers who have investigated: Zajac, Cholewa, Poprzecki, Waskiewicz, Langfort 2009 Mar 1, 45–50.

114 These benefits have even: Siegler and Hirscher from the department of sport, health, and exercise science, University of Hull, UK, conducted a study to observe "the ergogenic potential of sodium bicarbonate (NaHCO3) ingestion on

boxing performance." Ten amateur boxers were prematched for weight and boxing ability, and ingested either bicarbonate or a placebo. Sparring bouts consisted of four 3-minute rounds, each separated by a 1-minute rest. The paper concluded that a standard dose of bicarbonate "improves punch efficacy during 4 rounds of sparring performance." Siegler JC, Hirscher K. Sodium bicarbonate ingestion and boxing performance. *J Strength Cond Res.* 2010 Jan;24(1):103–8.

116 Marathon runners are: Almond CS, Shin AY, Fortescue EB, et al. Hyponatremia among runners in the Boston Marathon. *N Eng J Med.* 2005 Apr 14;352(15):1550–6.

116 In a 2002 study: Ibid.

116 The state medical: Smith S. Marathon runner's death linked to excessive fluid intake. *Boston Globe.* August 13, 2002. www.remembercynthia.com/Hyponatremia_BostonGlobe.htm (accessed September 2, 2014).

116 Commenting on the tragedy: *Doctors:* Marathoner died from too much water, hyponatremia a danger in long-distance sports. WCVB 5. August 13, 2002. www.wcvb.com/Doctors-Marathoner-Died-From-Too-Much-Water/11334548#!bOn5pH (accessed September 2, 2014).

117 In his book *Facing Up*: Grylls B. *Facing Up: A Remarkable Journey to the Summit of Everest.* London: Pan; 2001:29.

118 Almost half of those: Maggiorini M. Mountaineering and altitude sickness. *Ther Umsch.* 2001 Jun;58(6):387–93.

119 At least one study shows: In a dissertation by Dr. Zubieta-Calleja, entitled "Human adaptation to high altitude and to sea level," the author noted that "patients with high hematocrit values had nearly twice as long breath holding times as normal and were able to sustain desaturation (of oxygen) at very low levels." Zubieta-Calleja G. *Human Adaptation to High Altitude and to Sea Level: Acid-Base Equilibrium, Ventilation and Circulation in Chronic Hypoxia.* Copenhagen: VDM; 2010.

119 The air in mountainous: Gallagher SA, Hackett PH. High-altitude illness. *Emerg Med Clin North Am.* 2004 May;22(2):329–55.

120 Other symptoms arising: Hackett PH, Roach RC. High-altitude illness. *N Engl J Med.* 2001 Jul 12;345(2):107–14.

120 This is a common occurrence: Moloney E, O'Sullivan S, Hogan T, Poulter LW, Burke CM. Airway dehydration: A therapeutic target in asthma? *Chest.* 2002 Jun;121(6):1806–11.

7. Bring the Mountain to You

121 World-renowned Brazilian: Lee F. Breathe right and win. Viewzone.com. www.viewzone.com/breathing.html (accessed August 15, 2012).

121 De Oliveira's goal was to: Ibid.

121 De Oliveira's techniques: Tom Piszkin. Interview with Luiz de Oliveira. Personal e-mail to Patrick McKeown, November 2012.

122 In total, the athletes: Ibid.

122 "But if you use my drill": Lee, Breathe right and win.

122 Maintaining form during: Ibid.

122 According to de Oliveira: Tom Piszkin. Interview with Luiz de Oliveira. Personal e-mail to Patrick McKeown, November 2012.

122 By the end of 1984: Joaquim Cruz. Wikipedia. en.wikipedia.org/wiki/Joaquim_Cruz (accessed April 20, 2013).

122 The legendary Czech: Litsky F. Emil Zatopek, 78, ungainly running star, dies. *New York Times.* November 23, 2000. www.nytimes.com/2000/11/23/sports/emil-zatopek-78-ungainly-running-star-dies.html (accessed September 2, 2014).

122 On the first day: Vaughan D. "Running": A great Czech athlete inspires a French novelist. Radio Praha. August 24, 2013. www.radio.cz/en/section/books/running-a-great-czech-athlete-inspires-a-french-novelist (accessed September 2, 2014).

122 On the second day: Ibid.

123 Rupp's headphones: Fairbourn J. Farah "confused" when making 2 hour claim says Salazar. *Eightlane* (blog). October 6, 2013. eightlane.org/farah-confused-making-2-hour-claim-salazar/ (accessed September 2, 2014).

131 However, unlike some: Sheila Taormina. Personal e-mail to Patrick McKeown, December 9, 2013.

131 After breath-hold training: French researcher Lemaître found that breath holds could also improve swimming coordination. After breath-hold training, swimmers showed increases in VO_2 peak as well as an increase in the distance traveled with each swimming stroke. The researchers concluded that their studies indicated that "breath-hold training improves effectiveness at both peak exercise and submaximal exercise and can also improve swimming technique by promoting greater propulsive continuity." Lemaître F, Seifert L, Polin D, Juge J, Tourny-Chollet C, Chollet D. Apnea training effects on swimming coordination. *J Strength Cond Res.* 2009 Sep;23(6):1909–14.

132 Researchers investigating: In addition to studying the effects of breath-hold training on swimming coordination, Lemaître and colleagues also investigated the effects of short repeated breath holds on breathing pattern in trained underwater hockey players (UHP) and untrained subjects (controls). Twenty male subjects were recruited, with ten members of a national underwater hockey team allocated to the UHP group, and ten subjects with little training and no breath hold experience allocated to the control group.

The subjects performed 5 breath holds while treading water with their faces immersed. The breath holds were spaced 5 minutes apart and performed after a deep but not maximal inhalation. The underwater hockey players were noted to have reduced breathlessness and higher concentration of CO_2 in exhaled breath after the test (ETCO2). Lemaître et al. 2007 Nov–Dec, 407–14.

132 In addition, lactate: Ibid.

132 The effect of this method: Researchers from the Human Performance Laboratory, University of Calgary, Canada, conducted a study to investigate the relationship between a decrease of oxygen concentration during exercise and erythropoietin (EPO) production. Five athletes cycled for 3 minutes at an intensity greater than maximal (supramaximal) at two different elevations: 1,000m and 2,100m. Oxygen saturation of hemoglobin was lower than 91 percent for approximately 24 seconds during exercise at 1,000 meters and for 136 seconds during exercise at 2,100 meters, with EPO levels increasing by 24 percent and 36 percent respectively following the exercise. Roberts D, Smith DJ, Donnelly S, Simard S. Plasma-volume contraction and exercise-induced hypoxaemia modulate erythropoietin production in healthy humans. *Clin Sci.* 2000 Jan;98(1):39–45.

Korean researchers Choi et al. carried out a study on 263 subjects to determine the relationship between hematocrit levels and obstructive sleep apnea (involuntary holding of the breath during sleep). Patients with severe sleep apnea had significantly higher levels of hematocrit than mild and moderate OSA. Study findings showed that hematocrit levels were significantly correlated with percent of time spent at oxygen saturation of below 90 percent, as well as average oxygen saturation. Choi JB, Loredo JS, Norman D, et al. Does obstructive sleep apnea increase hematocrit? *Sleep Breath.* 2006 Sep;10(3):155–60.

132 Lowering oxygen saturation: Roberts, Smith, Donnelly, Simard 2000, 39–45.

135 Maintaining an oxygen: Ibid.

136 Performing just 5 maximum: Lemaître et al. 2007 Nov–Dec, 407–14. Schagatay E, Haughey H, Reimers J. Speed of spleen volume changes evoked by serial apneas. *Eur J Appl Physiol.* 2005 Jan;93(4):447–52.

137 Breath-hold divers: Resting Hb mass in trained breath-hold divers was 5 percent higher than in untrained divers. In addition, breath-hold divers showed a larger relative increase to Hb after three apneas. The paper noted that "the long-term effect of apnea training on Hb mass might be implicated in elite divers' performance." Lemaître, Joulia, Chollet 2010, 413–5.

137 In addition, experienced: Matt Richardson investigated the hematological responses to maximal apneas performed by three groups: elite apneic divers, elite cross-country skiers, and untrained subjects. Pretest hemoglobin tended to be

higher in the diver group than both skiers and untrained individuals. Each subject was required to perform 3 maximal breath holds separated by 2 minutes of rest and normal breathing. Following the breath holds, all groups responded with increased hemoglobin, with divers showing the largest increase. The duration of the third breath-hold time was 187 seconds in divers, 111 seconds in skiers, and 121 seconds in untrained individuals. The authors observed that the higher Hb concentration in divers "suggests that regular apnea practice could impart a specific training effect, effecting haematological responses to apnea in a manner that differs from that of exercise training." Richardson M, de Bruijn R, Holmberg HC, Björklund G, Haughey H, Schagatay E. Increase of hemoglobin concentration after maximal apneas in divers, skiers, and untrained humans. *Can J Appl Physiol.* 2005 Jun;30(3):276-81.

Splenic size was measured before and after repetitive breath-hold dives to approximately 6 meters in ten Korean *ama* (diving women) and in three Japanese males who were not experienced in breath holding. Following the breath holds, splenic size and hematocrit were unchanged in the Japanese male divers. In the *ama,* splenic volume decreased 19.5 percent, hemoglobin increased by 9.5 percent, and hematocrit increased 9.5 percent. The study showed that long-term repeated apneas induce a stronger spleen contraction and resultant hematological response. Hurford WE, Hong SK, Park YS, et al. Splenic contraction during breath-hold diving in the Korean *ama. J Appl Physiol.* 1990 Sep;69(3):932–6.

138 For example, a study: Andersson and colleagues from Lund University in Sweden conducted a study involving fourteen healthy volunteers who performed a series of 5 maximal duration breath holds while their faces were immersed in water. The authors observed that breath-hold time increased by 43 percent across the series of breath holds. Andersson JP, Schagatay E. Repeated apneas do not affect the hypercapnic ventilatory response in the short term. *Eur J Appl Physiol.* 2009 Mar;105(4):569–74.

138 Another study found: French researchers Joulia et al. observed that trained divers who had 7–10 years of experience in breath-hold diving were able to hold their breath for up to 440 seconds at rest, compared with inexperienced individuals who held their breath for 145 seconds at most. Joulia F, Steinberg JG, Wolff F, Gavarry O, Jammes Y. Reduced oxidative stress and blood lactic acidosis in trained breath-hold human divers. *Respir Physiol Neurobiol.* 2002 Oct;133(1–2):121–30.

138 Similarly, the duration: Joulia F, Steinberg JG, Faucher M, et al. Breath-hold training of humans reduces oxidative stress and blood acidosis after static and dynamic apnea. *Respir Physiol Neurobiol.* 2003 Aug 14;137(1):19–27.

8. Finding the Zone

141 "Your whole being is involved": Geirland J. Go with the flow. *Wired.* September 1996.

143 Recounting the race: Bentley R, Langford R. *Inner Speed Secrets: Mental Strategies to Maximize Your Racing Performance.* Osceola, WI: MBI Pub. Co.; 2000.

143 We no longer give: Kevin Kelly. Personal e-mail to Patrick McKeown, August 15, 2013.

144 Selker suggests: Turning into digital goldfish. BBC. February 22, 2002. news.bbc.co.uk/2/hi/1834682.stm (accessed September 2, 2014).

144 Reading a piece: Bilton N. Steve Jobs was a low-tech parent. *New York Times.* September 10, 2014. www.nytimes.com/2014/09/11/fashion/steve-jobs-apple-was-a-low-tech-parent.html?_r=0 (accessed January 24, 2015).

144 "They haven't used it": Ibid.

145 "The focus on oneself": Giggsy doing it for himself. Yahoo Eurosport UK. November 28, 2013. Available at sg.newshub.org/giggsy_doing_it_for_himself_53525.html (accessed September 2, 2014).

145 Earl Woods believed: Carter B. Tiger emerges from Woods as golfing icon. ESPN Classic. espn.go.com/classic/biography/s/woods_tiger.html (accessed September 2, 2014).

145 In the film: *The Legend of Bagger Vance* movie review. Movieguide. www.movieguide.org/reviews/the-legend-of-bagger-vance.html (accessed September 2, 2014).

146 In an interview: Isaacson W. *Steve Jobs.* CD. Simon & Schuster Audio; 2011.

146 Eight Marine infantry: Johnson DC, Thom N, Stanley E, et al. Modifying resilience mechanisms in at-risk individuals: A controlled study of mindfulness training in marines preparing for deployment. *Am J Psychiatry.* 2014 Aug;171(8):844–53.

146 In other studies with: Hruby P. Marines expanding use of meditation training. *Washington Times.* December 5, 2012. www.washingtontimes.com/news/2012/dec/5/marines-expanding-use-of-meditation-training (accessed December 3, 2014).

147 Until recently: Congleton C, Hölzel BK, Lazar SW. Mindfulness can literally change your brain. *Harvard Business Review.* January 8, 2015. hbr.org/2015/01/mindfulness-can-literally-change-your-brain (accessed January 24, 2015).

147 A team of scientists: Ibid.

162 However, in a: *ROG—The Ronan O'Gara Documentry* [sic]. RTE Television. www.rte.ie/tv/programmes/rog.html (accessed September 2, 2014).

167 A study investigating: The results showed that hyperventilation significantly affects mental performance. Bruno Balke and colleagues from the U.S. Air Force

School of Aviation at Randolph Field, Texas, researched the effect of hyperventilation among jet pilots and whether it was a possible cause of unexplainable aircraft accidents. The objective of the study was to investigate the affect of hyperventilation on muscular activity that required mental processing. Six healthy male individuals were tested on a U.S. Air Force coordination apparatus before, during, and after hyperventilation of 30 minutes duration. Lung carbon dioxide decreased to 12–15 mmHg during, hyperventilation (normal $PaCO_2$ is 40 mmHg). The researchers found that mental performance deteriorated by 15 percent when the concentration of arterial carbon dioxide reduced to 20 to 25 mmHg, and by 30 percent when carbon dioxide concentration in arterial blood lowered to 14 mmHg. Balke B, Lillehei JP. Effect of hyperventilation on performance. *J Appl Physiol.* 1956 Nov 1;9(3):371–4.

167 Another study found: Researchers from the department of psychology, University of Leuven, Belgium, investigated the effect of reduced carbon dioxide on performance that required attention. The paper reported that hyperventilation that reduces arterial concentration of carbon dioxide is associated with physiological changes in the brain and with subjective symptoms of dizziness and concentration problems. The researchers found that more errors were made and progressively slower reaction times were observed during recovery from lower pressure of carbon dioxide. Van Diest I, Stegen K, Van de Woestijne KP, Schippers N, Van den Bergh O. Hyperventilation and attention: Effects of hypocapnia on performance in a stroop task. *Biol Psychol.* 2000 Jul;53(2–3):233–52.

167 A study from the department: Ley and colleagues from the department of psychology and statistics at the University at Albany in New York found that students with high anxiety had lower levels of end-tidal carbon dioxide and faster respiration frequency than low-anxiety students. The study found that the "high-test-anxiety group reported a greater frequency of symptoms of hyperventilation and a larger drop in level of end-tidal CO_2 during testing than low-test-anxiety group." Ley R, Yelich G. Fractional end-tidal CO_2 as an index of the effects of stress on math performance and verbal memory of test-anxious adolescents. *Biol Psychol.* 2006;Mar;71(3):350–1.

168 Sleep apnea: Kim et al. 2010, 533–9. Kreivi et al. 2010, 488–94. Ohki et al. 1996, 228–30. Lee et al. 2007 Jun, 1102–6. Scharf, Cohen 1998 Oct, 279–87; quiz 287–90. Wasilewska, Kaczmarski 2010, 201–12. Rappai, Collop, Kemp, deShazo 2003, 2309–23. Izu SC, Itamoto CH, Pradella-Hallinan M, et al. Obstructive sleep apnea syndrome (OSAS) in mouth breathing children. *Braz J Otorhinolaryngol.* 2010 Sep–Oct;76(5):552–6.

9. Rapid Weight Loss Without Dieting

179 Sherpas and others: Ghose T. Altitude causes weight loss without exercise. *Wired.* February 4, 2010. www.wired.com/wiredscience/2010/02/high-altitude-weight-loss (accessed August 1, 2013).

179 Based on this observation: Wasse LK, Sunderland C, King JA, Batterham RL, Stensel DJ. Influence of rest and exercise at a simulated altitude of 4,000 m on appetite, energy intake, and plasma concentrations of acylated ghrelin and peptide YY. *J Appl Physiol.* 2012 Feb;112(4):552–9. Kayser B, Verges S. Hypoxia, energy balance and obesity: From pathophysiological mechanisms to new treatment strategies. *Obes Rev.* 2013 Jul;14(7):579–92. Lippl FJ, Neubauer S, Schipfer S, et al. Hypobaric hypoxia causes body weight reduction in obese subjects. *Obesity (Silver Spring).* 2010 Apr;18(4):675–81. Westerterp-Plantenga MS, Westerterp KR, Rubbens M, Verwegen CR, Richelet JP, Gardette B. Appetite at "high altitude" [Operation Everest III (Comex-'97)]: A simulated ascent of Mount Everest. *J Appl Physiol.* 1999 Jul;87(1):391–9. Pugh, LGCE. Physiological and medical aspects of the Himalayan Scientific and Mountaineering Expedition, 1960–61. *Br Med. J.* 1962 Sep 8;2(5305):621–7. Rose MS, Houston CS, Fulco CS, Coates G, Sutton JR, Cymerman A. Operation Everest II: Nutrition and body composition. *J. Appl. Physiol.* 1988 Dec;65(6):2545–51.

179 In tests with mice: Ling Q, Sailan W, Ran J, et al. The effect of intermittent hypoxia on bodyweight, serum glucose and cholesterol in obesity mice. *Pak J Biol Sci.* 2008 Mar 15;11(6):869–75.

179 Researchers concluded: Qin L, Xiang Y, Song Z, Jing R, Hu C, Howard ST. Erythropoietin as a possible mechanism for the effects of intermittent hypoxia on bodyweight, serum glucose and leptin in mice. *Regul Pept.* 2010 Dec 10;165(2–3):168–73.

179 Of course, living: Kayser, Verges 2013 Jul, 579–92.

180 Incorporating both aerobic: Mercola J. Do shorter, higher intensity workouts for better results with the Peak 8 Fitness Interval Training Chart. Mercola.com. fitness.mercola.com/sites/fitness/Peak-8-fitness-interval-training-chart.aspx (accessed August 1, 2013). Rev up your workout with interval training. Mayo Clinic. www.mayoclinic.com/health/interval-training/SM00110 (accessed August 1, 2013).

180 is well documented: Ng DM, Jeffery RW. Relationships between perceived stress and health behaviors in a sample of working adults. *Health Psychol.* 2003 Nov;22(6):638–42. Epel E, Lapidus R, McEwen B, Brownell K. Stress may add bite to appetite in women: A laboratory study of stress-induced cortisol and eating behavior. *Psychoneuroendocrinology.* 2001 Jan;26(1):37–49.

Oliver G, Wardle J, Gibson EL. Stress and food choice: A laboratory study. *Psychosom Med.* 2000 Nov–Dec;62(6):853–65. Grunberg NE, Straub RO. The role of gender and taste class in the effects of stress on eating. *Health Psychol.* 1992;11(2):97–100.

180 relationship issues: Wheeler C. Eliminate emotional overeating and shed unwanted pounds. Mercola.com. May 20, 2006.articles.mercola.com/sites/articles/archive/2006/05/20/eliminate-emotional-overeating-and-shed-unwanted-pounds.aspx (accessed August 1, 2013).

180 Results showed that high: Ng, Jeffery 2003; 638–42.

181 These practices of meditation: Tapper K, Shaw C, Ilsley J, Hill AJ, Bond FW, Moore L. Exploratory randomised controlled trial of a mindfulness-based weight loss intervention for women. *Appetite.* 2009 Apr;52(2):396–404. Hepworth NS. A mindful eating group as an adjunct to individual treatment for eating disorders: A pilot study. *Eat Disord.* 2011 Jan–Feb;19(1):6–16. Kristeller JL, Hallett CB. An exploratory study of a meditation-based intervention for binge eating disorder. *J Health Psychol.* 1999 May;4(3):357–63. Dalen J, Smith BW, Shelley BM, Sloan AL, Leahigh L, Begay D. Pilot study: Mindful Eating and Living (MEAL): Weight, eating behavior, and psychological outcomes associated with a mindfulness-based intervention for people with obesity. *Complement Ther Med.* 2010 Dec;18(6):260–4.

181 Controlling stress: Wing RR, Phelan S. Long-term weight loss maintenance. *Am J Clin Nutr.* 2005 Jul;82(1 Suppl):222–5.

10. Reduce Physical Injury and Fatigue

186 To investigate the relationship: Oxford University Press. Famous performers and sportsmen tend to have shorter lives. *ScienceDaily.* April 17, 2013. www.sciencedaily.com/releases/2013/04/130417223631.htm (accessed September 2, 2014).

187 premature aging: Gruber J, Schaffer S, Halliwell B. The mitochondrial free radical theory of ageing—where do we stand? *Front Biosci.* 2008 May 1;13:6554–79.

187 damage to the heart: Patil HR, O'Keefe JH, Lavie CJ, Magalski A, Vogel RA, McCullough PA. Cardiovascular damage resulting from chronic excessive endurance exercise. *Mo Med.* 2012 Jul–Aug;109(4):312–21.

187 dementia: Bennett S, Grant MM, Aldred S. Oxidative stress in vascular dementia and Alzheimer's disease: A common pathology. *J Alzheimers Dis.* 2009;17(2):245–57.

187 Free radicals are: Devasagayam TP, Tilak JC, Boloor KK, Sane KS, Ghaskadbi SS, Lele RD. Free radicals and antioxidants in human health: Current status and future prospects. *J Assoc Physicians India.* 2004 Oct;52:794–804.

187 During physical exercise: Urso ML, Clarkson PM. Oxidative stress, exercise, and antioxidant supplementation. *Toxicology.* 2003 Jul 15;189(1–2):41–54. Powers SK, Jackson MJ. Exercise-induced oxidative stress: Cellular mechanisms and impact on muscle force production. *Physiol Rev.* 2008 Oct;88(4):1243–76. Finaud, Lac, Filaire E 2006, 327–58.

187 Investigations into physical: Powers SK, Nelson WB, Hudson MB. Exercise-induced oxidative stress in humans: Cause and consequences. *Free Radic Biol Med.* 2011 Sep 1;51(5):942–50. Kanter M. Free radicals, exercise and antioxidant supplementation. *Proc Nutr Soc.* 1998 Feb;57(1):9–13. A study by Jackson from the department of medicine in the University of Liverpool noted that 30 minutes of excessive muscular activity in rats resulted in increased free radical activity. The researchers suggested that this phenomenon might play a role in causing muscle damage. Jackson MJ. Reactive oxygen species and redox-regulation of skeletal muscle adaptations to exercise. *Philos Trans R Soc Lond B Biol Sci.* 2005 Dec 29;360(1464):2285–91. Machefer G, Groussard C, Rannou-Bekono F, et al. Extreme running competition decreases blood antioxidant defense capacity. *J Am Coll Nutr.* 2004 Aug;23(4):358–64.

Researchers at the department of medicine at the University of Helsinki in Finland conducted a study to determine the effects of physical training on free radical production. Nine fit male subjects were studied before and after three months of running and were found to have significantly decreased levels of all circulating antioxidants except for ascorbate during training. The conclusion reached was that "relatively intense aerobic training decreases circulating antioxidant concentrations." Bergholm R, Mäkimattila S, Valkonen M, et al. Intense physical training decreases circulating antioxidants and endothelium-dependent vasodilatation in vivo. *Atherosclerosis.* 1999 Aug;145(2):341–9.

188 Blood samples were: Machefer G, Groussard C, Rannou-Bekono F, et al. 2004, 358–64.

188 At first glance: Clarkson PM. Antioxidants and physical performance. *Crit Rev Food Sci Nutr.* 1995 Jan;35(1–2):131–41. Clarkson PM, Thompson HS. Antioxidants: What role do they play in physical activity and health? *Am J Clin Nutr.* 2000 Aug;72(2 Suppl):637–46. Urso, Clarkson 2003 Jul 15, 41–54. Sacheck JM, Blumberg JB. Role of vitamin E and oxidative stress in exercise. *Nutrition.* 2001 Oct;17(10):809–14.

188 Research has shown that: A paper published in the *Journal of Respiratory Physiology and Neurobiology* reported on a three-month breath-hold program that was superimposed onto the regular training of triathletes. The researchers found that by incorporating breath holding into physical exercise, "blood acidosis was reduced and the oxidative stress no more occurred." The paper concluded that

"these results suggest that the practice of breath-holding improves the tolerance to hypoxemia (inadequate level of oxygen in the blood) independently from any genetic factor." Joulia et al. 2003, 19–27.

Another study tested whether repeated breath holds by elite breath-hold divers to reduce oxygen pressure in the blood could result in reduced blood acidosis and oxidative stress. Trained divers with seven to ten years of experience in breath hold diving, and with an ability to hold their breath for up to 440 seconds during rest, were compared with a second group of non-divers who had at most a 145 second breath-hold time.

Both groups performed a breath hold during rest, followed by 2 minutes of forearm exercises during which the diver group performed a breath hold and the second group breathed as normal. Interestingly, the group who breathed as normal showed an increase in blood lactic acid concentration and oxidative stress. In the diver group, the changes in both lactic acid and oxidative stress were markedly reduced after both breath holds and exercise. The paper concluded that humans who are involved in a long-term training program of breath-hold diving have reduced blood acidosis and oxidative stress following breath holds and exercise. Joulia, Steinberg, Wolff, Gavarry, Jammes 2002, 121–30.

For those of you who might be concerned that reducing the effects of free radicals only relates to elite breath-hold divers, let me resolve your fears with the results of one final study. A 2008 paper published in the journal *Medicine & Science in Sports & Exercise* investigated the effects of breath holding on oxidative stress using two groups of people: a group of trained divers and a group of people with no diving experience at all. Results showed significant improvements in antioxidant activity across both groups, with little difference between the divers and non-divers. Bulmer AC, Coombes JS, Sharman JE, Stewart IB. Effects of maximal static apnea on antioxidant defenses in trained free divers. *Med Sci Sports Exerc.* 2008 Jul;40(7):1307–13.

189 Athletes with long: Joulia, Steinberg, Wolff, Gavarry, Jammes 2002, 121–30.

189 Research spanning thirty: Fisher-Wellman K, Bloomer RJ. Acute exercise and oxidative stress: A 30 year history. *Dyn Med.* 2009 Jan 13;8:1.

189 Exercising several times: Radak Z, Chung HY, Goto S. Systemic adaptation to oxidative challenge induced by regular exercise. *Free Radic Biol Med.* 2008 Jan 15;44(2):153–9. Campbell PT, Gross MD, Potter JD, et al. Effect of exercise on oxidative stress: A 12-month randomized, controlled trial. *Med Sci Sports Exerc.* 2010 Aug;42(8):1448–53. Majerczak J, Rychlik B, Grzelak A, et al. Effect of 5-week moderate intensity endurance training on the oxidative stress, muscle specific uncoupling protein (UCP3) and superoxide dismutase (SOD2)

contents in vastus lateralis of young, healthy men. *J Physiol Pharmacol.* 2010 Dec;61(6):743–51.

189 More rigorous training: Finaud, Lac, Filaire 2006, 327–358.

189 Studies show that: Shing CM, Peake JM, Ahern SM, et al. The effect of consecutive days of exercise on markers of oxidative stress. *Appl Physiol Nutr Metab.* 2007 Aug; 32(4):677–85. Gomez-Cabrera MC, Domenech E, Viña J. Moderate exercise is an antioxidant: Upregulation of antioxidant genes by training. *Free Radic Biol Med.* 2008 Jan 15;44(2):126–31.

191 The naked mole rat: Veselá A, Wilhelm J. The role of carbon dioxide in free radical reactions of the organism. *Physiol Res.* 2002;51(4):335–9.

191 This might also explain: Buffenstein R. Negligible senescence in the longest living rodent, the naked mole-rat: Insights from a successfully aging species. *J Comp Physiol B.* 2008 May;178(4):439–45. Veselá, Wilhelm 2002, 335–9.

191 develop cancer: Rathi A. Cancer immunity of strange underground rat revealed. Conversation. June 19, 2013. theconversation.com/cancer-immunity-of-strange-underground-rat-revealed-15358 (accessed September 2, 2014).

191 "Even when scientists": Ibid.

191 Although a few days' rest: Researchers in the United States investigated the effects of detraining in collegiate competitive swimmers who commonly take a month off from training following a major competition. The study measured aerobic fitness, resting metabolism, mood state, and blood lipids in each swimmer during two tests: one in a trained state, and another after a resting period of five weeks. The results of the second test clearly showed an increase of body weight, fat mass, and waist circumference, and a decrease of VO_2 peak. The authors suggested, therefore, that coaches and athletes ought to be aware of the negative consequences of detraining from swimming. Ormsbee MJ, Arciero PJ. Detraining increases body fat and weight and decreases VO2peak and metabolic rate. *J Strength Cond Res.* 2012 Aug; 26(8):2087–95.

Koutedakis Y. Seasonal variation in fitness parameters in competitive athletes. *Sports Med.* 1995 Jun;19(6):373–92.

A study of senior rugby league players found that a period of six weeks of inactivity produced a significant decrease in VO_2 max. Allen GD. Physiological and metabolic changes with six weeks detraining. *Aust J Sci Med Sport.* 1989;21(1): 4–9. Godfrey RJ, Ingham SA, Pedlar CR, Whyte GP. The detraining and retraining of an elite rower: A case study. *J Sci Med Sport.* 2005 Sep;8(3):314–20. Mujika I, Padilla S. Detraining: Loss of training-induced physiological and performance adaptations. Part II: Long term insufficient training stimulus. *Sports Med.* 2000 Sep;30(3):145–54.

192 For some, high-intensity: Toumi H, Best T. The inflammatory response: Friend or enemy for muscle injury? *Br J Sports Med.* 2003 Aug;37(4):284–6.

11. Improve Oxygenation of Your Heart

193 The same tragedy: Go AS, Mozaffarian D, Roger VL, et al. Heart disease and stroke statistics—2014 update: A report from the American Heart Association. *Circulation.* 2014 Jan 21;129(3):e28–e292.

194 "Isn't it the irony": Ringertz N. Alfred Nobel's health and his interest in medicine. Nobelprize.org. December 1, 1998. www.nobelprize.org/alfred_nobel/biographical/articles/ringertz (accessed September 2, 2014).

194 In 1896 Alfred Nobel: Ibid.

194 In an ironic twist: The Nobel Prize in Physiology or Medicine 1998. NobelPrize.org. www.nobelprize.org/nobel_prizes/medicine/laureates/1998 (accessed September 2, 2014).

195 Sometimes referred: Chang, *Nitric Oxide, the Mighty Molecule,* 201. *Dr Louis Ignarro on nitric oxide 1.* www.youtube.com/watch?v=FsA04n2k6xY (accessed September 2, 2014). Ignarro, *NO more heart disease;* 2006.

195 Nitric oxide sends: Ibid.

195 If the blood clots: Ibid. Ignarro, *NO more heart disease;* 2006. *Dr Louis Ignarro on nitric oxide 2.* www.youtube.com/watch?v=B4KHlP8Bttw (accessed September 2, 2014).

195 Nitric oxide plays: Ibid.

195 According to Nobel: *Dr Louis Ignarro on nitric oxide 2.*

195 As we breathe in through the nose: Lundberg, Weitzberg 1999, 947–52.

196 Dr. David Anderson: Breathe deep to lower blood pressure, doc says. Associated Press. July 31, 2006. Available at www.nbcnews.com/id/14122841/ns/health-heart_health/t/breathe-deep-lower-blood-pressure-doc-says (accessed September 2, 2014).

196 A plausible explanation: Mourya M, Mahajan AS, Singh NP, Jain AK. Effect of slow- and fast-breathing exercises on autonomic functions in patients with essential hypertension. *J Altern Complement Med.* 2009 Jul;15(7):711–7. Pramanik T, Sharma HO, Mishra S, Mishra A, Prajapati R, Singh S. Immediate effect of slow pace bhastrika pranayama on blood pressure and heart rate. *J Altern Complement Med.* 2009 Mar;15(3):293–5.

196 But the middle path: Goto C, Higashi Y, Kimura M, et al. Effect of different intensities of exercise on endothelium-dependent vasodilation in humans: Role of endothelium-dependent nitric oxide and oxidative stress. *Circulation.* 2003 Aug 5;108(5):530–5.

197 In conclusion, the researchers: University of Exeter. Beetroot juice boosts stamina, new study shows. *ScienceDaily.* August 7, 2009. www.sciencedaily.com/releases/2009/08/090806141520.htm (accessed September 2, 2014).

197 In 1909, American: Dr. Henderson, 70, physiologist, dies; Director of Yale Laboratory, expert on gases, devised methods of revival *New York Times.* February 20, 1944. Henderson Y. Acapnia and shock: I. Carbon dioxide as a factor in the regulation of the heart rate. *Amer Jour Phys.* 1908 Feb;21(1):126–56.

197 In a paper entitled: Henderson 1908 Feb, 126–56.

198 The one thing: Lum 1975, 375–83.

200 This state of hypocapnia: Rutherford JJ, Clutton-Brock TH, Parkes MJ. Hypocapnia reduces the T wave of the electrocardiogram in normal human subjects. *Am J Physiol Regul Integr Comp Physiol.* 2005 July;289(1):R148–55. Hashimoto K, Okazaki K, Okutsu Y. The effect of hypocapnia and hypercapnia on myocardial oxygen tension in hemorrhaged dogs. *Masui.* 1990 Apr;39(4):437–41. Kazmaier S, Weyland A, Buhre W, et al. Effects of respiratory alkalosis and acidosis on myocardial blood flow and metabolism in patients with coronary artery disease. *Anesthesiology.* 1998 Oct;89(4):831–7. Neill WA, Hattenhauer M. Impairment of myocardial O_2 supply due to hyperventilation. *Circulation.* 1975 Nov;52(5):854–8.

200 Since low levels: Neill, Hattenhauer 1975, 854–8.

200 On March 2, 2004: Cormac Trust. www.thecormactrust.com (accessed December 12, 2012).

200 Tributes to Cormac: Ibid.

201 In search of the reasons: Dr. Domenico Corrado from the department of cardiac, thoracic, and vascular sciences at the University of Padvoa, Italy, presented to the 2009 European Society of Cardiology congress in Barcelona. The title of his presentation was "Electrical repolarization changes in young athletes: What is abnormal?" Dr. Corrado recognized that ECG changes in athletes are common and usually reflect remodeling of the heart as an adaptation to regular physical training. However, although an abnormal ECG reading of T-wave inversion is rarely observed in healthy athletes, it was found to be a potential expression of an underlying heart disease, presenting a risk of sudden death from cardiac arrest during sport. Corrado D. Electrical repolarization changes in young athletes: What is abnormal? ESC Congress 2009. Barcelona. August 31, 2009. spo.escardio.org/eslides/view.aspx?eevtid=33&id=2616 (accessed April 15, 2013).

In a 2008 paper published in the *New England Journal of Medicine,* researchers examined a database of 12,550 trained athletes. From this, a total of 81 athletes who had no apparent cardiac disease were identified as having ECG abnormalities of deeply inverted T-waves. Of the 81 athletes with abnormal ECGs,

1 died suddenly at the age of 24 years from cardiac failure. Of the 80 surviving athletes, 3 developed heart disease at the ages of 27, 32, and 50, including 1 who had an aborted cardiac arrest. The researchers concluded that markedly abnormal ECGs in young and apparently healthy athletes may represent the initial expression of underlying cardiac disease, and that athletes with such ECG patterns merit continued clinical surveillance. Pelliccia A, Di Paolo FM, Quattrini FM, et al. Outcomes in athletes with marked ECG repolarization abnormalities. *N Eng J Med.* 2008 Jan 10;358:152–61.

Laukkanen and colleagues from the University of Kuopio, Finland, investigated the association between ST depression and the risk of sudden cardiac death in a population-based sample of 1,769 men. During the eighteen years of follow-up, a total of 72 deaths occurred due to sudden cardiac death in those found with asymptomatic ST segment depression. The risk of sudden cardiac death was found to have increased among men with asymptomatic ST segment depression during exercise and during the recovery period. It was noted "asymptomatic ST-segment depression was a very strong predictor of sudden cardiac death in men with any conventional risk factor but no previously diagnosed coronary heart disease." Laukkanen JA, Mäkikallio TH, Rauramaa R, Kurl S. Asymptomatic ST-segment depression during exercise testing and the risk of sudden cardiac death in middle-aged men: A population-based follow-up study. *Eur Heart J.* 2009 Mar;30(5):558–65.

201 When this happens: Kasper DL, Harrison TR. *Harrison's Principles of Internal Medicine.* New York: McGraw-Hill Medical Publishing Division; 2005.

201 In assessing ECG: Thompson PD. Exercise and the heart: The good, the bad, and the ugly. *Dialog Cardiovasc Med.* 2002;7(3):143–62.

201 Studies have found: Corrado. Electrical repolarization changes in young athletes.

201 However, certain abnormal: Ibid.

201 Markedly abnormal: Pelliccia, Di Paolo, Quattrini, et al. 2008 Jan 10, 152–61.

202 ST segment depression: Kligfield P, Lauer MS. Exercise electrocardiogram testing beyond the ST segment. Circulation. 2006;114:2070–82.

202 In a study including: Laukkanen, Mäkikallio, Rauramaa, Kurl 2009, 558–65.

202 A study conducted by: Alexopoulos D, Christodoulou J, Toulgaridis T, et al. Repolarization abnormalities with prolonged hyperventilation in apparently healthy subjects: Incidence, mechanisms and affecting factors. *Eur Heart J.* 1996 Sep;17(9):1432–7.

204 Dr. Lum was well known: Laurence Claude Lum. Royal College of Physicians. munksroll.rc plondon.ac.uk/Biography/Details/6079 (accessed September 2, 2014).

204 Dr. Lum dedicated: Ibid.

204 Both activities increase breathing: Rutherford, Clutton-Brock, Parkes 2005, R148–55. Hashimoto, Okazaki, Okutsu 1990, 437–41. Kazmaier et al. 1998, 831–7. Neill, Hattenhauer 1975, 854–8.

205 Up to 10 percent: Chelmowski MK, Keelan MH Jr. Hyperventilation and myocardial infarction. *Chest.* 1988 May;93(5): 1095–6.

205 In one particular study: Ibid.

205 A study of twenty patients: Elborn JS, Riley M, Stanford CF, Nicholls DP. The effects of flosequinan on submaximal exercise in patients with chronic cardiac failure. *Br J Clin Pharmacol.* 1990 May;29(5):519–24.

206 This research: Buller NP, Poole-Wilson PA. Mechanism of the increased ventilatory response to exercise in patients with chronic heart failure. *Br Heart J.* 1990 May;63(5):281–3. The authors observed that patients with breathing problems had reduced arterial carbon dioxide and increased breathing volume per minute. Furthermore, patients with problem breathing had greater impaired cardiac function. Fanfulla FM, Mortara A, Maestri R, et al. The development of hyperventilation in patients with chronic heart failure and Cheyne-Stokes respiration: A possible role of chronic hypoxia. *Chest.* 1998 Oct;114(4):1083–90. Vasiliauskas D, Jasiukeviciene L. Impact of a correct breathing stereotype on pulmonary minute ventilation, blood gases and acid-base balance in post–myocardial infarction patients. *Eur J Cardiovasc Prev Rehabil.* 2004 Jun;11(3):223–7.

206 In a 2004 study published: Vasiliauskas, Jasiukeviciene 2004 June, 223–7.

206 Based on improvements: Ibid.

206 Other studies confirm: Patients who practiced breathing exercises for reversing chronic hyperventilation evidenced significantly higher carbon dioxide levels and lower respiratory rates when compared with pretreatment levels measured three years earlier. The authors concluded, "Breathing retraining has lasting effects on respiratory physiology, and is highly correlated with a reduction in reported functional cardiac symptoms." DeGuire S, Gevirtz R, Kawahara Y, Maguire W. Hyperventilation syndrome and the assessment of treatment for functional cardiac symptoms. *Am J Cardiol.* 1992 Sep 1;70(6):673–7.

207 We know that hyperventilation: Researchers from the division of cardiology, Kumamoto University School of Medicine, Japan, investigated the hyperventilation test as a clinical tool to induce coronary artery spasm (narrowing of blood vessels to the heart). The study involved 206 patients with coronary spasm and 183 patients without angina at rest (nonspasm). Each patient performed hyperventilation for 6 minutes. Of the spasm group, 127 showed positive responses to the test, including electrocardiographic changes attributable to reduced blood flow. No one in the nonspasm group showed any ischemia (narrowing of blood

flow). When clinical characteristics were compared, high disease activity and severe arrhythmias were significantly higher in the hyperventilation test positive patients than in the negative patients (69 percent vs. 20 percent). The authors concluded that "hyperventilation is a highly specific test for the diagnosis of coronary artery spasm, and that hyperventilation test-positive patients are likely to have life-threatening arrhythmias during attacks." Nakao K, Ohgushi M, Yoshimura M, et al. Hyperventilation as a specific test for diagnosis of coronary artery spasm. *Am J Cardiol.* 1997 Sep 1;80(5):545–9.

207 but studies have: In the aptly titled paper "Death by Hyperventilation: A Common and Life-Threatening Problem During Cardiopulmonary Resuscitation," researchers tested the hypothesis that excessive ventilation rates during the performance of CPR by overzealous but well-trained rescue personnel increases the likelihood of death. The paper investigated thirteen adult deaths where manual CPR with an average of 30 breaths per minute was administered to patients. The paper also documented a study investigating ventilation per minute and survival rate during cardiac arrest in pigs. Three groups of seven pigs were treated with 12 breaths, 30 breaths, or 30 breaths plus carbon dioxide per minute. Survival rates in the groups were as follows: six out of seven pigs treated with 12 breaths per minute, one out of seven pigs treated with 30 breaths per minute, and one out of seven pigs treated with 30 breaths per minute plus carbon dioxide. The authors commented that "despite seemingly adequate training, professional rescuers consistently hyperventilated patients during out-of-hospital CPR," and that "additional education of CPR providers is urgently needed to reduce these newly identified and deadly consequences of hyperventilation during CPR." Aufderheide TP, Lurie KG. Death by hyperventilation: A common and life-threatening problem during cardiopulmonary resuscitation. *Crit Care Med.* 2004 Sep;32(9 Suppl):345–51.

In a paper entitled "Do We Hyperventilate Cardiac Arrest Patients?" published in the journal *Resuscitation* in 2007, researchers studied data from twelve patients who had received manual ventilation by a self-inflating bag in the emergency department of a UK hospital. Results showed that the number of manual breaths administered to the patients varied from 9 to 41 per minute, with an average of 26. The corresponding median volume of air per minute was 13 liters. The researchers noted that while guidelines on the number of breaths to administer during CPR are well known, "it would appear that in practice they are not being observed." O'Neill JF, Deakin CD. Do we hyperventilate cardiac arrest patients? *Resuscitation.* 2007 Apr;73(1):82–5.

207 Researchers investigated instances: Ibid.

207 One study concluded: Ibid.

12. Eliminate Exercise-Induced Asthma

210 Exercise-induced asthma: Rundell KW, Im J, Mayers LB, Wilber RL, Szmedra L, Schmitz HR. Self-reported symptoms and exercise-induced asthma in the elite athlete. *Med Sci Sports Exerc.* 2001 Feb;33(2):208–13.

210 Interestingly, one study: Sidiropoulou MP, Kokaridas DG, Giagazoglou PF, Karadonas MI, Fotiadou EG. Incidence of exercise-induced asthma in adolescent athletes under different training and environmental conditions. *J Strength Cond Res.* 2012 Jun;26(6):1644–50.

211 Tackling asthma from: Zinatulin SN. *Healthy Breathing: Advanced Techniques.* Novosibirsk, Russia: Dinamika Publishing House; 2003.

212 Normal breathing volume: Johnson BD, Scanlon PD, Beck KC. Regulation of ventilatory capacity during exercise in asthmatics. *J Appl Physiol.* 1995 Sep;79(3):892–901. Chalupa DC, Morrow PE, Oberdörster G, Utell MJ, Frampton MW. Ultrafine particle deposition in subjects with asthma. *Environ Health Perspect.* 2004 Jun;112(8):879–82. Bowler SD, Green A, Mitchell CA. Buteyko breathing techniques in asthma: A blinded randomised controlled trial. *Med J Aust.* 1998 Dec 7–21;169(11–12):575–8.

212 of air per minute: Pulmonary structure and function. In: McArdle, Katch, Katch, *Exercise Physiology,* 263.

212 During an exacerbation of asthma: *GINA Report, Global Strategy for Asthma Management and Prevention.* Global Initiative for Asthma; 2014:74. www.gin asthma .org/guidelines-gina-report-global-strategy-for-asthma.html (accessed December 27, 2012).

213 A study at the Mater Hospital: Bowler, Green, Mitchell 1998, 575–8.

214 The reason for this: Ibid.

214 Further studies reinforced: McHugh P, Aitcheson F, Duncan B, Houghton F. Buteyko Breathing Technique for asthma: An effective intervention. *N Z Med J.* 2003 Dec 12;116(1187):U710. Cowie RL, Conley DP, Underwood MF, Reader PG. A randomised controlled trial of the Buteyko technique as an adjunct to conventional management of asthma. *Resp Med.* 2008 May;102(5);726–32.

214 Based on the fact: Ibid. Bowler, Green, Mitchell 1998, 575–8.

215 People diagnosed with asthma: Hallani M, Wheatley JR, Amis TC. Initiating oral breathing in response to nasal loading: Asthmatics versus healthy subjects. *Eur Respir J.* 2008 Apr;31(4):800–6. A paper published in the medical journal *Chest* noted that "asthmatics may have an increased tendency to switch to oral (mouth) breathing, a factor that may contribute to the pathogenesis of their asthma." Kairaitis K, Garlick SR, Wheatley JR, Amis TC. Route of breathing in patients with asthma. *Chest.* 1999 Dec;116(6):1646–52.

215 Air taken in through: Fried (ed.), *Hyperventilation Syndrome,* 1987.

215 The mouth is simply: Ibid.

216 Unlike nasal breathing: Djupesland PG, Chatkin JM, Qian W, Haight JS. Nitric oxide in the nasal airway: A new dimension in otorhinolaryngology. *Am J Otolaryngol.* 2001 Jan–Feb;22(1):19–32. Scadding G. Nitric oxide in the airways. *Curr Opin Otolaryngol Head Neck Surg.* 2007 Aug;15(4):258–63. Vural C, Güngör A. Nitric oxide and the upper airways: Recent discoveries. *Tidsskr Nor Laegeforen.* 2003 Jan;10(1):39–44.

216 Taking all these factors: Hallani M, Wheatley JR, Amis TC. Enforced mouth breathing decreases lung function in mild asthmatics. *Respirology.* 2008 Jun;13(4):553–8.

216 The paper concluded: Shturman-Ellstein R, Zeballos RJ, Buckley JM, Souhrada JF. The beneficial effect of nasal breathing on exercise-induced bronchoconstriction. *Am Rev Respir Dis.* 1978 Jul;118(1):65–73.

216 In simple terms: Researchers studied the effects of nasal breathing and oral breathing on exercise-induced asthma. Fifteen people were recruited for the study and asked to breathe only through their nose. The study found that "the post-exercise bronchoconstrictive response was markedly reduced as compared with the response obtained by oral (mouth) breathing during exercise, indicating a beneficial effect of nasal breathing." Mangla PK, Menon MP. Effect of nasal and oral breathing on exercise-induced asthma. *Clin Allergy.* 1981 Sep;11(5):433–9.

217 The difference between: In the words of respiratory consultant Dr. Peter Donnelly, which were published in the medical journal the *Lancet,* "In most land based forms of exercise, patterns of breathing are not constrained, ventilation increases proportionately throughout exercise and end tidal CO_2 tensions are either normal or low. Therefore, there is no hypercapnic (increased carbon dioxide) stimulus for bronchodilation (airway opening) and asthmatics have no protection." Donnelly PM. Exercise induced asthma: The protective role of CO_2 during swimming. *Lancet.* 1991 Jan 19;337(8734):):179–80.

217 At the beginning of this chapter: Sidiropoulou, Kokaridas, Giagazoglou, Karadonas, Fotiadou 2012 Jun, 1644–50.

217 Although the act of: Uyan ZS, Carraro S, Piacentini G, Baraldi E. Swimming pool, respiratory health, and childhood asthma: Should we change our beliefs? *Pediatr Pulmonol.* 2009 Jan;44(1):31–7. Fjellbirkeland L, Gulsvik A, Walløe A. Swimming-induced asthma. *Tidsskr Nor Laegeforen.* 1995 Jun 30;115(17):2051–3. Bernard A, Carbonnelle S, Michel O, et al. Lung hyperpermeability and asthma prevalence in schoolchildren: Unexpected associations with the at-

tendance at indoor chlorinated swimming pools. *Occup Environ Med.* 2003 Jun;60(6):385–94. Nickmilder M, Bernard A. Ecological association between childhood asthma and availability of indoor chlorinated swimming pools in Europe. *Occup Environ Med.* 2007 Jan;64(1):37–46.

13. Athletic Endeavor—Nature or Nurture?

221 In 1704: Cooper C. The Stud: Why retirement will be a full-time job for Frankel. *Independent.* October 26, 2012. www.independent.co.uk/sport/racing/the-stud-why-retirement-will-be-a-fulltime-job-for-frankel-8228820.html (accessed June 10, 2013).

221 Geneticist Patrick Cunningham: Cunningham EP, Dooley JJ, Splan RK, Bradley DG. Microsatellite diversity, pedigree relatedness and the contributions of founder lineages to thoroughbred horses. *Anim Genet.* 2001 Dec;32(6):360–4.

222 Although the natural: Abreu RR, Rocha RL, Lamounier JA, Guerra AF. Prevalence of mouth breathing among children. *J Pediatr (Rio J).* 2008;84(5):467–70.

223 It has been well: Tourne LP. The long face syndrome and impairment of the nasopharyngeal airway. *Angle Orthod.* 1990 Fall;60(3):167–76. Deb U, Bandyopadhyay SN. Care of nasal airway to prevent orthodontic problems in children. *J Indian Med Assoc.* 2007 Nov;105(11):640, 642. Harari D, Redlich M, Miri S, Hamud T, Gross M. The effect of mouth breathing versus nasal breathing on dentofacial and craniofacial development in orthodontic patients. *Laryngoscope.* 2010 Oct;120(10):2089–93.

224 Yogi Bhajan: Bhajan. The Living Chronicles of Yogi Bhajan aka the Siri Singh Sahib of Sikh Dharma. WhoAreTheSikhs.com. www.harisingh.com/LifeAccordingToYogiBhajan.htm.

224 The ancient Buddhist: Mallinson J. *The Khecarîvidyâ of Adinathâ.* London and New York: Routledge; 2007:17–19.

225 In a paper written by researchers: Wong EM, Ormiston ME, Haselhuhn MP. A face only an investor could love: CEOs' facial structure predicts their firms' financial performance. *Psychol Sci.* 2011 Dec;22(12):1478–83.

225 In a separate study: Ibid.

226 Chronic, habitual mouth: Okuro RT, Morcillo AM, Sakano E, Schivinski CI, Ribeiro MÂ, Ribeiro JD. Exercise capacity, respiratory mechanics and posture in mouth breathers. *Braz J Otorhinolaryngol.* 2011 Sep–Oct;77(5):656–62. Okuro RT, Morcillo AM, Ribeiro MÂ, Sakano E, Conti PB, Ribeiro JD. Mouth breathing and forward head posture: Effects on respiratory biomechanics and

exercise capacity in children. *J Bras Pneumol.* 2011 Jul–Aug;37(4):471–9. Conti, Sakano, Ribeiro, Schivinski, Ribeiro 2011, 471–9.

226 "Many of these children": Jefferson Y. Mouth breathing: Adverse effects on facial growth, health, academics and behavior. *Gen Dent.* 2010 Jan–Feb;58(1):18–25.

226 Dr. Egil Peter Harvold: Harvold EP, Tomer BS, Vargervik K, Chierici G. Primate experiments on oral respiration. *Am J Orthod.* 1981 Apr;79(4):359–72. Miller AJ, Vargervik K, Chierici G. Sequential neuromuscular changes in rhesus monkeys during the initial adaptation to oral respiration. *Am J Orthod.* 1982 Feb;81(2):99–107. Moses AJ. Airways and appliances. *CDS Rev.* 1989 Mar;82(2):50–7. Available at: www.tmjchicago.com/uploads/airwaysandappliances.pdf (accessed September 2, 2014).

227 Dr. Harvold's: Vargervik K. Egil Peter Harvold, Orthodontics: San Francisco. Calisphere, University of California. texts.cdlib.org/view?docId=hb0h4n99rb&doc.view=frames&chunk.id=div00029&toc.depth=1&toc.id= (accessed September 2, 2014).

227 Research has suggested: Trabalon M, Schaal B. It takes a mouth to eat and a nose to breathe: Abnormal oral respiration affects neonates' oral competence and systemic adaptation. *Int J Pediatr.* 2012;2012:207605. O'Hehir T, Francis A. Mouth vs. nasal breathing. *Hygientown.* September 2012. www.hygienetown.com/hygienetown/article.aspx?i=297&aid=4026 (accessed September 2, 2014).

229 Development of the lower jaw: Meridith HV. Growth in head width during the first twelve years of life. *Pediatrics.* 1953 Oct;12(4):411–29.

230 The detrimental effects: Schreiner C. Nasal airway obstruction in children and secondary dental deformities. University of Texas Medical Branch, Department of Otolaryngology, Grand Rounds Presentation. 1996.

14. Exercise as if Your Life Depends on It

232 Dozens of studies: Blair SN, Cheng Y, Holder JS. Is physical activity or physical fitness more important in defining health benefits? *Med Sci Sports Exerc.* 2001 Jun;33(6 Suppl):379–99. Crespo CJ, Palmieri MR, Perdomo RP, et al. The relationship of physical activity and body weight with all-cause mortality: Results from the Puerto Rico Heart Health Program. *Ann Epidemiol.* 2002 Nov;12(8):543–52. Oguma Y, Sesso HD, Paffenbarger RS Jr, Lee IM. Physical activity and all cause mortality in women: A review of the evidence. *Br J Sports Med.* 2002 Jun;36(3):162–72.

232 Not only this: A most interesting study investigating the relationship between regular physical exercise and cardiovascular health was conducted as far back as 1952 by Scottish epidemiologist Dr. Jeremy Morris. Commonly, known as the

"bus conductor study," Dr. Morris and colleagues investigated the incidence of heart attacks across 31,000 male transport workers between the ages of 35 and 65 who worked during the years 1949 and 1950. Morris JN, Heady JA, Raffle PA, Roberts CG, Parks JW. Coronary heart-disease and physical activity of work. *Lancet* 1953 Nov 21;265(6795):1053–7.

232 The same study: Andrade J, Ignaszewski A. Exercise and the heart: A review of the early studies, in memory of Dr R.S. Paffenbarger. *B C Med J.* 2007 Dec;49(10):540–6.

Appendix: Upper Limits and Safety of Breath Holding

290 Another effect is bradycardia: Lindholm P, Lundgren CE. The physiology and pathophysiology of human breath-hold diving. *J Appl Physiol.* 2009 Jan;106(1):284-92. Espersen, Frandsen, Lorentzen, Kanstrup, Christensen 2002 May, 2071–9.

290 Breaking point: Lin YC, Lally DA, Moore TO, Hong SK. Physiological and conventional breath-hold break points. *J Appl Physiol.* 1974 Sep;37(3):291–6.

290 While it is extremely: Nunn JF. *Applied Respiratory Physiology.* London and Boston: Butterworths; 1987.

291 Research by Ivancev: Ivancev et al. investigated whether repetitive breath holding blunts the chemoreceptors, resulting in reduced reactivity to carbon dioxide. Blunted chemoreceptors are recognized as a common result of obstructive sleep apnea. Ivancev et al. tested the hypothesis that repeated breath holds, which are an integral part of breath-hold diving, blunt cerebrovascular reactivity to hypercapnia. Two groups of seven elite breath-hold divers and seven non-divers were involved in the test. The study noted that breath-hold divers had a greater tolerance to carbon dioxide, largely the result of lower breathing frequency. The findings of the study were "that the regulation of the cerebral circulation in response to hypercapnia is intact in elite breath-hold divers, potentially as a protective mechanism against the chronic intermittent cerebral hypoxia and/or hypercapnia that occurs during breath-hold diving." Therefore, regular breath-hold practice does not impair cerebrovascular reactivity to high carbon dioxide pressure. Ivancev V, Palada I, Valic Z, et al. Cerebrovascular reactivity to hypercapnia is unimpaired in breath-hold divers. *J Physiol.* 2007 Jul 15;582(Pt 2):723–30.

291 A further study by Joulia: With repeated practice, elite breath-hold divers are able to sustain very long breath holds that induce a severe drop in oxygen without causing brain injury or blackouts. A study of the circulatory effects of apnea in elite breath-hold divers by Joulia et al. showed that bradycardia and periph-

eral vasoconstriction were accentuated in breath-hold divers compared with non-divers. In addition, a decrease in oxygen saturation was less and carotid arteries blood flow was greater among the breath-hold divers during apnea. Joulia F, Lemaître F, Fontanari P, Mille ML, Barthelemy P. Circulatory effects of apnea in elite breath-hold divers. *Acta Physiol (Oxf).* 2009 Sep;197(1):75–82.

293 "Please do not practice": Preparation. Navy Seals. www.navyseals.com/preparation (accessed August 20, 2012).

Acknowledgments

This book would not have been possible without the spirited support of Oxygen Advantage coaches and athletes and gifted advice from a number of extraordinary people. In particular, I would like to thank my book agent, Doug Abrams, and his team, including Lara Love, who spurred me on to rewrite the entire original manuscript, providing insights, clarity, and editing to help ensure that I communicated my message just "as if I were talking to a fellow down at the pub." To Cassie Jones and the publishing team at William Morrow, thank you so much for your dedicated support and commitment in making this book a reality. Thank you, Claudia Connal and colleagues from Piatkus and Little Brown Book Group, for getting this book into the hands of readers throughout Europe, South Africa, Australia, and New Zealand.

Also helping me along this journey was Jo Gatford, who weaved her magic through 95,000 words, knitting the final manuscript into a readable and digestible form.

To my colleagues, including Dr. Alan Ruth, whose attention to detail astounds me, Tom Herron, Eoin Burns, Carol Baglia, Don Gordon, Eugenia Malyshev, Dr. Charles Florendo, Tom Piszkin, and William L. Robbins—thank you so much for dropping your work to read through the manuscript, providing honest and direct feedback on what needed to be changed and where. James O'Toole, Eamon Howley, and Danny

Dreyer—thank you for providing encouragement and belief that the message imparted in this book would be valuable to athletes.

Thank you to Sarah Gallagher for her expertise in checking papers, adding studies, and clarifying others to ensure that I interpreted the information on cardiovascular health accurately. Special thanks to Dr. Joseph Mercola, who has devoted his life's work to educating the public about simple, safe, and effective preventative measures to ensure good health. Dr. Mercola, the world needs more people like you!

To soccer coach Don O'Riordan and the dedicated players of Galway WFC, thank you for your assistance in tailoring the Oxygen Advantage program to a team sport environment.

I must also thank my illustrator, Rebecca Burgess, who always comes out on top with beautiful images that communicate the content visually.

Thanks also to my wife, Sinead, who, with her great sense of humor, insisted that my acknowledgment to her shouldn't be as boring as my usual ones. Well, you can be the judge of that! To my beautiful daughter, Lauren, may you grow up as a nasal breather.

And finally, thank you to my readers for taking a chance on me by buying and reading this book. I hope you derive lifelong benefits from the words within.

About the Author

PATRICK MCKEOWN was educated at Trinity College, Dublin, and later studied under the auspices of the founder of the Buteyko Method, the late Dr. Konstantin Buteyko. From a very young age, Patrick suffered from asthma and relied on an array of medicines and inhalers until he discovered the Buteyko Method at age twenty-six. By applying the principles from Dr. Buteyko, Patrick experienced immediate relief from his asthmatic symptoms—a feat that more than twenty years of medication had failed to accomplish. Learning to breathe through his nose and reducing his breathing volume positively changed his life in so many ways.

For the first time ever, Patrick's asthma was under control.

This life-changing discovery motivated Patrick to change his career in order to help children and adults who suffered from similar breathing problems. Following his time in Russia in 2002, Patrick has been teaching the Oxygen Advantage program in Australia, the United States, and throughout Europe. He has written seven books, including in their respective fields three Amazon.com bestsellers, and has been invited to speak at dental and respiratory conferences around the world.

In collaboration with the University of Limerick, Patrick was the instructor in a clinical study investigating the Buteyko Method as a treatment for rhinitis in asthma. As published in the leading rhinitis journal *Otolaryngology,* the results showed a 70 percent reduction of nasal symptoms in participants, including snoring, loss of smell, nasal congestion, and difficulty breathing through the nose.

The Oxygen Advantage program is based on Patrick's experience of working with thousands of clients and hundreds of health care professionals, along with his extensive research on breath-hold training over the past thirteen years. A number of the exercises in this book were specifically created by Patrick to help athletes to improve their sports performance.

Patrick has been interviewed internationally for radio and television, including by Dr. Joseph Mercola, founder of the world's largest health website, Mercola.com. Patrick's message is simple: Breathe through your nose in a calm, gentle, and efficient manner. That's all that matters. Patrick's passion for his subject enables him to continue developing and refining his techniques to bring relief to asthma sufferers worldwide, and to improve the fitness and well-being of athletes and nonathletes alike.

Your profession is not what brings home your pay.
Your profession is what you were put on earth to do with
such passion and such intensity that it becomes spiritual in calling.

—VINCENT VAN GOGH

About OxygenAdvantage.com

Visit our website for app downloads,
coaching, Skype consultations, and webinars.

By now, I hope that you have begun to put the Oxygen Advantage program into practice. For some, the theory and exercises in this book will be enough to completely revolutionize their exercise routine and health. For others, practical experience may be needed to help them to apply their knowledge and understanding of the Oxygen Advantage program. As you make progress, you may wish to take advantage of the personal services of our expert Oxygen Advantage coaches or our online Skype course and webinar.

- Oxygen Advantage coaches are available in several locations throughout the United States, Europe, Australia, and New Zealand. Our coaches are specifically trained to help you make the best progress possible, safely and quickly.

- Skype courses offer the opportunity to benefit from a one-to-one consultation with the author. During the Skype course, you will receive expert guidance, feedback, a specifically tailored program, and the motivation you need to ensure you make optimal progress.

- Our Oxygen Advantage webinar shows Patrick teaching each of the exercises to athletes, allowing you to see how the techniques are practically applied in real time.

If you would like to find out more or get in touch, please visit OxygenAdvantage.com. Your feedback and suggestions are greatly appreciated.

Index

F

G

H

K

L

M

R

S